B. LAPP

Comfrey—Healer Through History

Twenty-four hundred years ago, Greek soldiers used Comfrey to staunch their wounds. Medieval herbalists discovered that it could heal ulcers. Today's dermatologists find it a boon for dozens of skin ailments; and vegetarians eat young tender Comfrey leaves as a unique source of vitamin B-12. The legendary healing power of Comfrey has found its scientific basis in modern times and its reputation is spreading rapidly. Ben Charles Harris, who wrote *Eat the Weeds* and *Ginseng*, has explored history, science and his own extensive experience to describe Comfrey as food and medicine and to explain its cultivation and methods of use.

BEN CHARLES HARRIS'S

Comfrey

Keats Publishing, Inc. New Canaan, Connecticut

Ben Charles Harris's
COMFREY: WHAT YOU NEED TO KNOW

Pivot Original Health Edition published 1982
Copyright © 1982 by Faye Harris
Special contents copyright © 1982 by Keats Publishing, Inc.

Printed in the United States of America

ISBN: 0-87983-160-X
Library of Congress Catalog Card Number: 82-80699

Keats Publishing, Inc.
27 Pine Street, New Canaan, Connecticut 06840

CONTENTS

ABOUT BEN CHARLES HARRIS

An appreciation from the publishers

Ben Charles Harris became interested in herbs through his herbalist grandfather and during his lifetime was recognized as one of America's leading authorities on herbs. After working nearly forty years as pharmacist in his drugstore in Worcester, Massachusetts, he devoted his entire career to herbology and health with all the enthusiasm of a Culpeper. He wrote a dozen or so books, including the bestseller *Eat the Weeds*, taught classes on herbs and nutrition at Worcester State College and the Natural History Museum in Worcester (where he was curator of Economic Botany for twenty-two years), and conducted his own radio program "Yours for Better Health." He was also an ardent lecturer in this hemisphere on herbs and their culinary and medicinal uses; he conducted field trips as readily as he found interested participants; and for years he wrote a column for *Herald of Health* Magazine.

His knowledge of natural health and nutrition led him into typical active involvement: he held positions as President of the Boston Nutrition Society and of the Boston Chapter of the Natural Hygiene Society and as Vice-President of the Massachusetts chapter of the Natural Food Associates.

Ben Harris probably read every book ever written by a herbalist, but his conclusions were drawn from his own pragmatic experience, research and experiment. A visit to his kitchen revealed a variety of rootings, bakings, dryings, fermentings and steepings, indicating the curiosity of the moment, all carefully noted, to be compared with other findings and recorded.

He died in February, 1978. His pleasure in his work never flagged, and an extraordinary number of pupils and followers succumbed to the contagion of that pleasure, to his humor and to the delight of his presence which is sadly missed.

Keats Publishing, Inc.
New Canaan, Connecticut
April, 1982

CHAPTER 1

COMFREY:
A HEALER
THROUGHOUT
HISTORY

The healing arts have used Comfrey almost constantly for nearly twenty-five hundred years. An Old World plant native to Asia, this purposeful, health-restoring and wound-healing herb has been in demand in many lands and by all social classes. It has appeared in the writings of poets and its therapeutic efficacy soundly praised throughout the centuries by herbalists, country folk and physicians.

The word Comfrey comes from the Latin *con firma*, alluding to the uniting of bones, and from the Latin *confervere*, to boil or grow together, or to heal. Throughout history the reparative properties of plants have influenced their naming. Consound is a name given to Bugle, *Ajuga reptans*; Daisy, *Bellis perennis*; and Larkspur, *Delphinium consolida.* It comes from the Latin and French, meaning sound or whole. Note, too, Goldenrod's generic name, *solidago* (solid = whole; ago = I create).

Comfrey is a leading member of the Borage family which features as the major noticeable characteristics bluish flowers and bristly hairs. It is of the genus Symphytum, and that name, from the Greek, means coming together. The common Comfrey is S. officinale, denoting its important place in the herbalist's shop. With an average height of three feet, it blooms from June to September, producing flowers ranging from creamy white to rose to bluish purple.

Beginnings of recorded use

By 400 B.C. the plant we know as Comfrey was already in use in Greece. Herodotus, that nation's illustrious historian, recorded its use and recommended it to staunch severe bleeding mechanically. The Greeks later used the root to cure bronchial problems. In his herbal *Alexiphar-mica*, Nicander, Greek poet-physician of the second century B.C., mentions the plant as a remedy for poisons. The famous Greek physician Galen (A.D. 130–200) mentions its worthiness in his writings, too.

The Romans also appreciated the plant's remedial powers. Dioscorides, first century Greek physician in the employment of Nero as medical officer to the Roman legions, had prescribed the plant, which he called *pectos*, for its wound-healing and bone-knitting virtues. He, the Father of Materia Medica, on which homeopathy is based, had frequently used the herb to heal festering sores, insect bites, animal clawings, painful sandal blisters and armor abrasions. His contemporary, the Roman writer and naturalist Pliny, had observed: "The roots are so glutinative that they well solder or glew together meat that is chopt in pieces, seething in a pot, and make it into a lump. The same bruysed and layed in the manner of a plaster doth heal all fresh and green wounds." Pliny called the plant alum and solidago, which as noted above was also a synonym for Goldenrod. Other Roman herbalist-physicians employed this valuable healing agent while on expeditions throughout northern Europe.

The herb appears in monastery writings and herbals (leech-books) from A.D. 1000. "This Wort (plant) strengthens the man," we read in the Saxon Leechdom, a collection of recipes, prescriptions and remedies for various injuries. Other Saxon herbariums recommended it for those "bursten within" (possibly from internal bleedings, ruptures, hernias, etc.) for which purpose, to give one example, Comfrey leaves were heated in or over hot, near-ash embers, ground and stirred into honey, and then taken on an empty stomach.

The Benedictines, Cistercians and other religious orders need be credited with furthering the cultivation of Comfrey plants during long stretches of warfare and unavoidable neglect. Comfrey became one of the mainstay occupants of monastery gardens during the Middle Ages and for good reason: to heal the assortment of wounds and ruptures of returning soldiers.

Comfrey's qualities were well noted by herbalists during and following the Middle Ages, and by the late sixteenth century the herb had reached a high level of popularity, being used in most of the ways which will be discussed in this book. King and pauper alike cultivated it in their gardens. It served the country folk quite satisfactorily for a variety of ailments—as a vulnerary or remedy for healing external wounds or skin sores and as a highly touted remedy for broken bones and internal bleeding. A wild-growing species served John Gerard, herbalist supreme during Queen Elizabeth's reign, in these and other ways. He had noted that Comfrey healed "ulcers of the lungs and ulcers of long continuance. The rootes of Comfrey stamped and the juice drunke with wine helpeth those that spit blood, and healeth all inward wounds and burstings."

William Turner recommended the roots for external purposes: they served well "if they be broken and dronken for them that spitte blood and are bursten. The same layd to, are good to glew together freshe woundes. The rootes are also good to be layd to inflammation and especially of the fundament, with the leaves of Groundsell (Senecio vulgaris). "(*Herball*, 1568) / Dodoens, in *Cruydeboek*, 1578, furthered Turner's statement, adding that when "mingled with sugar, syrops or honny [the roots] are good to be layd upon all hoate tumours." The roots are similarly reported in Bulleyn's *Herbal* of 1562, in the *Adversaria* of Pena and Lobelius (1570), in the latter's *Stirpium Historia* and in the writings of Camerarius (1586). Gerard's *Herball* (1597) tells us the herb was most useful for "ulcers of the lunges and kidnies though they have been of long continuance." In the mid-seventeenth century, J. Bauhin would concur with these views, affirming that a decoction of Comfrey root was a must for all wounds, internal bleedings, and even for broken bones (*Historia*).

Enter master herbalist John Parkinson, apothecary to King James I. His widely read *Theater of Plants* (1640), the bible for contemporary herbalists and physicians alike, raised Comfrey to a new high of well-deserved acceptance. Parkinson noted various applications. Internally, the herb employed as an expectorant "helpeth the ulcers of the lungs causing the phlegme that oppressed them to be easily spit forthe." A water distilled presumably from the leaves and roots was intended "also to take away the fit of agues [fevers] and to allay the sharpness of humours." And a syrup, he found, was "effectual for all those inward griefs and hurts."

For external purposes, Parkinson cited his findings that the roots "being bruised and outwardly applied, helpeth sore, fresh wounds, [and] cuts immediately, and laid thereto, by glueing together their lips and [are] especially good for ruptures and broken bones." The cooked bruised roots were spread on leather and applied to a gouty area. In cases of hemorrhoids, a decoction (the concentrated liquid left after boiling down water and sliced plant root) would "cool the inflammation of the parts thereabouts and give ease of paines." Putrescent ulcers, gangrene "and the like" would be helped similarly by Comfrey. Like Pliny, Parkinson observed that "if boyled with dissevered pieces of flesh in a pot it [the root] will joyne them together againe." He further makes the horticultural contribution of directing that the plant must be grown in a "temperate degree, not as others say hot."

In her *Curious Herball*, Elizabeth Blackwell (*not* the Dr. Elizabeth Blackwell who was pioneer of women medical practitioners) mentions similar virtues attributed to Comfrey.

Comfrey in the modern age

New England Rarities Discovered was first printed in London in 1672 by a simplist (that is, his herbal was a book of single or simple herbs) named John Josselyn. His visits to Massachusetts (in 1638 for fifteen months, twenty-five years later for eight years) provided excellent opportunity to observe that "such garden herbs amongst us as do thrive here: such as Parsley, Carrots, Sage and Marigold and that many familiar plants as have sprung up since the English planted and kept cattle in New England." Good examples of the Englishman's herbal transplants were the well-known Plantain, Mallow, Nettles, Dandelion, Shepherd's Purse, Wormwood, Knotgrass and the hardy perennial "Compherie with the white flower."

Early settlers raised the herb in Salem and Plymouth, the Governor planted his on Boston Common, and homeowners of New Netherlands introduced it around 1649 to their gardens. The first frugal New Englanders avoided "frivolities" such as spending precious minutes to raise ornamental flowers, but Comfrey had already proven its worth as a plaster to external pus-laden sores and as a cough remedy. Thus it was for very sound reasons that this most hardy plant was dubbed "Comforty."

In his *Compleat Herbal* (1719) Townefort relates a number of "vertures" of Comfrey. He quotes Hieronymous Rensurerus: "a certain person [had been] cured of a malignant ulcer, pronounced to be a cancer by the surgeons and left by them as incurable, by applying twice a day the root of Comfrey bruised, having first peeled off the external blackish bark or rind; but the cancer was not above eight to ten weeks standing."

About sixty years later, a brilliant English physician, Dr. William Withering (who would copy a Shropshire woman herbalist's use of the garden-cultivated Foxglove for dropsy and later as a heart stimulant in cardiac decompensation) recorded various uses of Comfrey in his *Systematic Arrangement Of British Plants*, Vol.II, 1812. He lists its edibility, but notes that not all animals seek the herb as forage: "Sheep and goats eat it. Cows are not fond of it. Horses may refuse it." A most nutritious food, he concluded, for the herb "when burnt ... is said to afford a larger quantity of ashes than any other vegetable, often 1/7th of its weight." One sentence commands special attention: "Dr. Blair attributes a narcotic power to this plant."

Then came William Woodville who, in 1832, wrote of its healing properties: "The dried root, boiled in water, renders a large proportion of the fluid slimy; and the decoction inspissated, yields a strong flavorless mucilage similar to that obtained from Althea (Marsh Mallow) but somewhat stronger, equal-bodied or more tenacious and somewhat larger quantity, amounting to three quarters the weight of Comfrey." For this reason, he says, it was substituted for Althea, their emollient and dumulcent properties being quite equal. (*Medical Botanist*)

Other contemporary writers noted that a decoction of the roots was "esteemed in diarrhea, dysentery, blennorhagia [profuse mucous discharge, especially from the vagina], and pulmonary catarrh." It was "corrector of irritation of the intestines and mucous membranes". These writers alternately extolled and decried the leaves of Prickly Comfrey ("Trottles") as a forage plant. Three cuttings were taken in April, July and September. "Cattle eat the early shoots with great avidity. The plant does not communicate any disagreeable flavor to the milk." Farm folk, they observed, often partook of the stems "blanched like celery" during the winter months.

The mid-1800s saw an active resurgence in the use of Comfrey and other herbal preparations. Dr. A. Curtis noted its highly valued therapeutics in his *Botanico-Medical Recorder*, Vol. 6, 1838. He and other medical contemporaries would heed the urgings of Dr. Samuel Thomson, the most prominent of that period's eclectics, to undertake a complete discontinuance of chemical drugs, especially the much used, highly dangerous calomel, the "mineral poison," and to begin an immediate return to herbal medicines. In 1842, Dr. Morris Mattson had recommended (in *American Vegetable Practice*) Comfrey for many of the uses I have recorded. The herb's virtues are confirmed in the *Medical Botany* (1847) of Dr. R.E. Eglesfeld Griffith. Comfrey was recommended as a "demulcent and tonic" in the treatment of lung disease, as noted in the 1897 edition of *Lilly's Handbook of Pharmacy and Therapeutics*.

Moreover, as late as 1923, the herb was recognized in the *British Pharmaceutical Codex*. Its authors stated that Comfrey root from time immemorial had been employed as a domestic "simple" for applications to wounds, sores and ulcers of various kinds. The decoction not only is "applied to such wounds," it is given internally in gastralgia (stomach ache) and gastric ulcer in doses of 1/2 to 2 ounces three times a day.

Herb Comfrey had been included in *Squire's Companion to the British Pharmacopoeia*, 17th edition, which noted its astringent, mucilaginous and glutinous properties. Its author tells of an herbalist, actually a bone-setting expert, who gained fame through his expertise in treating bone fractures with Comfrey. His method: a pulp made of the scraped root was spread thickly on white cotton cloth or muslin which was then wrapped around the limb and made secure with bandage. In due time, a stiffness developed which gave immense strength and maintenance to the affected area. Not until the limb was well was the bandage removed.

Comfrey was officially recognized, Squire states, in the following Pharmacopoeias: in Belgium as *Radix Symphyti*; France, *Consoude*; Mexico, *Sinfito*; Portugal, *Consolida major*; Spain, *Sinfito major*. It appeared in the *British Pharmaceutical Codex*, in the 1923 edition. Though the herbals of the past few centuries may be considered quaint and outdated, they were the precursors of the modern "official" pharmacopoeia. Let us examine what we now know of its composition and its applications.

CHAPTER 2

THE CONSTITUENTS
AND WHAT
THEY DO

What can modern science tell us about Comfrey? What does the herb contain to bring the results observed over the centuries?

Analysis of Comfrey

First of all, we should note that its composition—and therefore its nutritional and its medicinal value—varies considerably as to type of plant, when picked, how dried or processed, and of course which part of the plant is employed and analyzed. The leaves, preferably the young shoots, may be eaten in a salad or as a vegetable or in soups; larger leaves are used in external applications; the root is often used for specific internal benefits. One widely available form of Comfrey is as a tea, usually made from the leaves.

Analysis of this form, commonly available in health stores, packaged singly or in blends, is a useful place to start:

Composition of Comfrey

Nutrient	Percentage
Protein	21.8 to 33.4
Fat	2.22
Carbohydrates	37.62
Crude fiber	9.38
Ash	15.06
Total digestible nutrients	86.5

Mineral analysis	Percentage
Iron	.016
Manganese	.0072
Calcium	1.700
Phosphorus	.820

Vitamins present	Milligrams per 100 grams
Thiamine	.5
Riboflavin	1.0
Nicotinic acid	5.0
Pantothenic acid	4.2
Vitamin B12	.07
Vitamin C	100.
Vitamin E	30.

Other	
Carotene	.170 parts per million
Allantoin	.18 milligrams per 100 grams

In a postscript to *Comfrey: An Ancient Medicinal Remedy*, Lawrence D. Hills indicates that the all-important minerals and vitamins will vary according to the season, as with other crops, and to various species. Here are his analyses of various Comfreys (using the leaves):

THE MINERAL ANALYSIS OF COMFREY

Bocking No. 4
Cut May 5

	percent
Calcium	2.35
Phosphoric acid	1.25
Potash	5.04
Iron	0.253
Manganese p.p.m.	137
Cobalt	trace

Bocking No. 15
Cut October 15

	percent
Calcium	2.38
Phosphoric acid	0.78
Potash	6.95
Iron	0.23
Manganese p.p.m.	133
Cobalt, less than 1 p.p.m.	

Bocking No. 14
Cut October 13

Calcium	2.77
Phosphoric acid	0.75
Potash	7.09
Iron	0.144
Cobalt less than 1 p.p.m.	

Bocking mixture

Calcium	2.58
Phosphoric acid	1.07
Potash	5.01
Iron	0.457
Manganese p.p.m.	201
Cobalt	trace

Symphytum officinale

Calcium	1.31
Phosphoric acid	0.72
Potash	3.09
Iron	0.098
Manganese p.p.m.	85

Mixed Bocking clones

	Potash percent	Phosphorous percent	Calcium percent
April 7	7.95	1.25	1.86
May 15	5.94	0.72	2.70
June 26	7.44	1.15	1.81
July 7	8.25	1.01	2.65
September 6	7.83	1.05	3.10

A British experimenter claims that the protein content figured on a freshweight basis of the Bocking mixture Comfrey cited above is 40 percent greater on June 11 (5.15 percent) than on August 30 (3.10 percent). It is enlightening to compare that Comfrey mixture with another type, the so-called "Russian" Comfrey (3.40 percent), cabbage (1.5 percent), kale (2.10), Alfalfa (4.10), ryegrass (2.90). As you can see, protein content is highest in the spring, although the percentage of allantoin, an important constituent which we will study in great detail, is lowest in that season.

Several of the constituents of Comfrey deserve and require special attention.

The herb offers four essential *amino acids*—tryptophan, lysine, isoleucine and methione. Except for isoleucine these are lacking in most vegetable proteins: this fact in addition to the high percentage of protein in Comfrey (24 percent of its dry matter) makes the plant of special interest to those striving to solve the problems of the world's hungry. Comfrey is in fact the fastest known builder of vegetable protein and provides seven times the protein found in soybeans.

Comfrey's high measure of *calcium*, with other herbal/vegetable minerals, helps to prevent and treat such disorders as muscle cramps, osteoporosis, fractures, etc.

Its *chlorophyll*, the life force of all green plants, needs little comment regarding its excellent healing powers in all situations—internal and external, restoring to health all diseased tissues, healing ulcers, burns and infections, detoxifying the blood stream, etc.

Compared with yeast, the average strain of Comfrey offers a minimum of four times as much *vitamin B12*. This anti-anemia vitamin, originally thought to be offered only in meat, fish and dairy products, has other sources: Sunflower seeds, soybean, Alfalfa, wheat germ, peanuts, certain root vegetables and sprouted seeds. Comfrey is, however, the only plant known to take vitamin B12 from the soil, explains Lawrence Hills. It thus should be of great usefulness in the vegetarian diet. Apart from its importance as builder of healthy blood, B12 is equally essential, wrote Earl W. Conroy in *Herald of Health* (July 1971), "in the healing of such diverse conditions as ulcers, arthritis, polio, muscular dystrophy, epilepsy, schizophrenia and mental confusion." (But that need not specifically indicate Comfrey's use in such cases.)

Its signaturing abundance of *mucilage* equals, if not surpasses, that of Elm bark and the Mallows, and indicates its like-cures-like importance:

the mucilage has the property of removing mucus or mucilage-like secretions from the bronchial or gastrointestinal linings. Besides, this jellyish exudation becomes a collodian-like first aid remedy and serves as a protective cover for damaged skin.

Among minor constituents of Comfrey are *starch and sugar* and its *asparagin* content (asparagin is an amino acid present also in Licorice, Marsh Mallow, asparagus and leguminous plants) which lets the herb act as a complementary diuretic.

We should take a good look at two other constituents which Heber W. Youngken, Professor Emeritus of the Massachusetts College of Pharmacy named *consolidine* and *symphtocynoglossine*. They are, he states in his *Textbook of Pharmacognacy*, "two poisonous alkaloids." Years later Dr. H.A. Vogel agreed that those substances, which are found mostly in the larger leaves, "can damage the central nervous system."

This factor has been long a subject of dispute among herb growers, herbalists, Comfrey users and myself. I have also long questioned the advisability of using such potentially dangerous herbs as Nightshade, Foxglove, Celandine et al., as well as the eating of boiled rhubarb or spinach, or fried foods (see page 64). The *U.S. Dispensatory* has warned that Wild Comfrey (Hounds Tongue), a close relative of Comfrey, contains a "poisonous alkaloid which acts upon the animal organism similarly to curare, an arrow poison used by South American Indians." The substance which paralyzes the central nervous system is a derivative of consolidin(e) which is found in both Hounds Tongue and Comfrey.

I feel strongly that one is perfectly secure in eating the early three- to five-inch leaves but the later larger leaves should be used for external purposes.

Publisher's Note: The debate and investigation has continued, prompting studies on the alkaloid content and toxicity in three research centers in England. These are reported in an appendix to Lawrence D. Hills's book, *Comfrey: Fodder, Food & Remedy* (Universe Books, 1976) by Dr. D.B. Long, who concludes: "the use of comfrey as a food for mankind or animals does not present a toxic hazard from alkaloids, there being no evidence of acute or chronic hepatic reactions either to the direct injection of purified alkaloid or to prolonged consumption of comfrey root flour, which has the highest alkaloid content, by rats."

A key constituent

Comfrey's most significant ingredient, however, is *allantoin*, a nitrogenous crystalline substance that appears as white crystals on the dried roots. It is produced mostly by the root system (it constitutes from 0.6 to 1 percent of that part) and much less by the terminal buds and the large leaves.

Allantoin is found in the bark of tree branches, sugar beets, French beans, peas, tobacco seeds, wheat sprouts, in germinating seeds, cauliflower leaves, mushrooms, soybeans, horsechestnut shoots and in liver.

It appears as a product of metabolism, too, the result of the alkaline oxidation of uric acids, in the urine of cattle and other grazing animals. (Its presence in animal urine in varying proportions affords a guide to veterinary diagnosis, similar to sugar in human diabetes.) Fetal fluid and the urine of pregnant women contain this substance. Minute quantities occur in the urine of healthy people which years ago was used as a handy antiseptic for cuts, scratches, etc.

In 1935, Dr. William Robinson of the U.S. Department of Agriculture observed that maggots, the larvae of certain flies, yielded this same substance and when placed on purulent open wounds and sores it afforded effective healing and rapid recovery. This was clearly in defiance of hospital hygiene but, as Andrew B. Devitt attests in the *Herb Grower Magazine*, "the cures were miraculous . . . Some of the most reluctant wounds responded beautifully."

The mucilaginous, allantoin-contained juice expressed from Comfrey (either as decoction, poultice or via the blender or juicer) has superb healing power when used externally: it "stimulates the growth of epithelium on ulcerated surfaces." (*Homeopathic Materia Medica*) Dr. R. W. Murray of Liverpool Hospital noted the same happening. He said that the liquid preparation would clean up the external surface "in a remarkable fashion." He also used Comfrey dressings in treating many men who were severely burned on the face, head and arms. Murray was so amazed by the highly satisfactory results that applications of the herb's wet dressings became standard procedure in the hospital.

It was through the discoveries and writings disclosed by an English physician, Dr. Charles J. Macalister, F.R.C.P., that Comfrey's usage as a therapeutic agent gained greater prominence. After years of personal investigations and many published studies, his findings regarding an "ancient medicinal remedy" resulted in his recording in 1936 the amazing and quick-healing attributes of this proliferant plant. Assorted tumors, pathological growths and malignancies, he wrote, were quickly and amazingly healed, much to the astonishment of his medical colleagues. Now Comfrey's therapeutic value was no longer an old wives' tale, no longer a scoffed-at folk remedy.

When his investigations led to his determining that Comfrey roots possessed large amounts of allantoin, the active principle "to which the virtues attributed to the Comfrey in *bygone* days are ascribable," he was supported by physicians and surgeons who had treated an assortment of external ulcers with root infusions, by clinicians whose patients did not respond to orthodox treatments, and by department heads at the University of Liverpool and their investigative organic chemists and biochemists.

The "father of Comfrey" had listened to the wisdom of the herb's past use (and users) and applied poultices to external tumors, and to others involving the mouth and nose. A face tumor, wrote Macalister, "completely disappeared from the face and there was no trace of it in the mouth. He [the patient] had no pain." Professor William Thompson, President of the Royal College of Surgeons in Ireland offered corroboration: "I am as satisfied as can be that the growth was malignant and of a bad type."

Dr. Macalister's experiments with plants ascertained allantoin's power as a cell proliferant. This is the prime factor that renders the herb a most valuable vulnerary and even more significantly an energetic generator of leucocytosis. And this condition in turn "helps to establish immunities in some infective conditions," remarked the doctor.

Of two earth-grown Hyacinths, one was well developed, with the beginning of a flower, and the other feeble, shorter and without sign of flower. Into the bulb of the latter, Macalister injected fifteen drops of 0.4 percent allantoin solution several times. The result: "Rapid growth ensued and it overgrew its more vigorous neighbor and flowered before it." Under the same conditions of temperature, soil and surroundings, he accomplished similar results with Tulips, Lily of the Valley and Chrysanthemum by injecting the allantoin solution into the base of the spikes of flowers.

His experiments led him to believe that allantoin can be "utilized by vegetable cells in connection with their proliferative processes, just as there seems to be proof that it has proliferative properties in connection with certain animal cells, if one may judge by the way it promotes healing in chronic and acute ulcerative conditions. . . ."

These same experiments may be conducted at home or in one's greenhouse. The use of allantoin in flower-growing could well develop into a horticulturist's long cherished dream. It may be possible to obtain results fairly comparable to such experiments by using a water soaking of the fresh or dried root and using the resultant liquid to water the plants. A decoction of the root is not advised since the allantoin is probably decomposed to some degree by the boiling process.

Allantoin in modern medicine

The pioneering stages are over. The various applications of allantoin in ointment, lotion or other vehicles have shown no discernible ill-effects. It has been described as "bland, stable, nontoxic and nonirritating." Now allantoin has respectability: it has been well accepted by the medical profession and associates and its virtues and healing efficacy highly lauded.

The *Merck Index of Chemicals and Drugs* says it is "used topically in suppurating wounds and resistant ulcers to stimulate the growth of healthy tissue; has also been given internally for gastric ulcer." Allantoin, states the *United States Dispensatory*, 32nd edition, had been similarly prescribed by British doctors during World War I for duodenal ulcers and for surface (skin) applications.

Podiatrists and skin specialists have long recognized allantoin's value in skin ulcers, open wounds, cuts and lacerations. Not only does it remove objectionable necrotic tissue; it strengthens epithelial formations and thus is indicated for internal and external ulceration. (Dr. D.W. Meixell and S.B. Mecca, *Journal of the American Podiatry Association,* Vol. 56:8)

F.R. Greenbaum had considered allantoin in "A New Granulation Substance with Especial Emphasis on Allantoin in Ointment Form," in which he observed that previous reports "demonstrate definitely the clinical value of allantoin ointment, and show we are dealing with a remarkable drug of great promise and wide possibilities in the treatment of infected or slowly healing wounds." (*American Journal of Surgery*, 34:259–65, 1936)

Ronald H. Johnson observed that "allantoin has been used many years by the medical profession to stimulate healthy granulation of tissues," that a number of preparations are to be obtained from its salts ("complexes"), and that its healing efficacy is lasting and such action non-irritating. (*Journal of the American Pharmaceutical Association*, November 1967)

Mr. Mecca's further remarks in "The Function and Applicability of the Allantoins" (Scientific Section of the Toilet Goods Association, May 1963) offer detailed descriptions of properties and uses of allantoin: it is soothing and emollient to calloused or chapped skin and if incorporated into creams, lotions and other cosmetic items greatly increases their effectiveness. Not only would dead or dying tissues be removed from the skin, the affected area would come alive with new tissue while the skin was continually being healed.

Dr. Joseph Kadaus, president of Bernadean University, Las Vegas, Nevada, has written that Swiss people, according to Dr. Alfred Vogel of Teufen, Switzerland, have long used a Comfrey poultice "to relieve the agonizing pains accompanying severe case of arthritis. By applying pulped Comfrey root to the painful parts, the pain will gradually fade out."

Dr. Kadaus also reported the removal of an ugly and troublesome wart with applications of Comfrey poultices. The large and nipple-shaped growth, located aside the nose, was removed surgically but then grew "bigger and more terrifying than ever." Poultices of the root were applied to the area and almost immediately "the nasty growth began to recede. In less than sixty days, it completely disappeared."

To rejuvenate aging skin, crow's feet and similar wrinkles Dr. Vogel suggests continuous applications of a preparation made by mixing a heaping teaspoonful of the powdered root into 3 or 4 ounces of an Irish Moss Lotion or lanolized cold cream. First apply warm water soaks to the face for several minutes, wipe dry, and thoroughly massage the preparation into the skin in the morning and at bedtime.

Lawrence D. Hills of the Henry Doubleday

Research Association in England states that this organization originated a Comfrey ointment. It contained an average 1 percent of allantoin in a coconut oil-lanolin base and produced, he said, "some quite remarkable cures of obstinate skin complaints ... With perseverence it will cure long standing and extensive psoriasis, and has been extremely effective against varicose ulcers of the feet, bad bed sores, bunions, shingles and a wide range of eczema cases." To use the ointment as a corn remedy, he suggests to "first cut the corn and then apply the ointment, covered with non-absorbent wool." It might be preferable to first hot-soak the area and ever so slightly pare the corn's peak.

Allantoin is included in externally applied preparations such as ointments for medical and veterinary purposes, lotions, creams, powders, gels, hand lotions, lipsticks, pomades, aftershave lotions, suppositories, suntan remedies (for its healing effect), scalp treatments, and athlete's foot and bromidosis powders.

In very recent years, these pharmaceutical products have been prescribed by medical doctors and have appeared in their *American Drug Index* and *Physicians Desk Reference*. The following list includes the physician-prescribed pharmaceuticals, the percentage of allantoin contained, and their indications, or uses:

Alphosyl Lotion, Cream and Gel, 2 percent; in the treatment of psoriasis and other chronic dermatoses with itching, scaling and erythema. Removes scales and crusts, lubricates the skin folds to lessen irritation. The lotion is indicated in psoriatic scalp skin conditions and suppresses itching and new scale formation and stimulates healing.

Anthryl, 2 percent; skin disorders.

AVC Dienestrol Cream, 2 percent; trichomoniasis, moniliasis and mixed vaginal infections.

AVC Improved Cream/ Suppositories; 2 percent in each, vaginitis, cervicitis, trichomoniasis and moniliasis.

Cutenal, 0.2 percent; general dermatologic purposes.

Diloderm, Cream/Foam contains an allantoin compound. Decreases edema, pruritus, and erythema.

Herpecin Lip Balm, an undeclared percentage of allantoin. "A soothing emollient cosmetically pleasant, lip balm. Relieves local dryness, soreness and irritation, cold sores, fever blisters and dry chapped lips."

Neo-Diloderm, beneficial in allergic dermatitis, eczemas, pruritis ani (rectal itch).

Masse-Cream has an undeclared percentage of allantoin. Massage for the nipple and areola of the breasts of pregnant and nursing women (remember the "old wives' tale" of applying a tepid herb poultice to heal such a condition?); a means of preventing or treating diaper rash; locally applied for bacteriostatic/healing effect.

Quada Cream, 0.4 percent; a "non-caine" antihistamine cream for the relief of itch and pain of contact dermatosis, due to poison ivy or oak, for pruritis ani.

Tenda Cream, 0.25 percent; a protective and therapeutic barrier cream for hand dermatitis and chapped skin due to soap and water (detergents) and soluble irritants.

Vagitrol Cream, 2 percent; vaginitis, cervicitis.

We have come full circle. In this century the scientific basis for the legendary healing power of Comfrey has been clarified. And modern science has taken the key constituent of the herb for its own. The health-minded individual can as well use Comfrey for its remarkable restorative effect.

CHAPTER 3

COMFREY AS A SOURCE OF NATURAL REMEDIES

The botanical discussion of Comfrey in Chapter 5 contains relevant information on the herb's invaluable effectiveness indicated by its external features. The reader might find these considerations helpful before we explore specific remedies.

AID FOR INTERNAL ILLS

Respiratory organs

Comfrey is demulcent and soothing to catarrhal and irritated conditions of colds, coughs and bronchial affections, and to the irritated mucous lining of the stomach and intestines.

Culpeper had suggested that a decoction of the root "heals inward hurts, wounds and ulcers of the lungs and causes phlegm that oppresses him to be easily spit forth." It gently stimulates and tones the mucous membranes of the respiratory organs. Best taken with Coltsfoot, Hoarhound and aromatics (Mint, Catnip, Marjoram et al.), Comfrey has a specific expectorant quality which increases the secretions of the bronchial linings and tones the relaxed part.

Another contributing factor in the singular healing phenomenon is more clearly demonstrated by noting that an aqueous extract of a variety of Comfrey, *Symphytum racemosus*, is effective against the *Staphylococcus aureus* organism.

Bronchial problems

Simmer 2 tablespoons each Comfrey root, Elecampane root, Hoarhound, Bloodroot, Coltsfoot in a quart of hot water for 1 hour.

NOTE: I recommend dried herbs throughout this book because the average reader has little or no access to the fluid extracts or tinctures recommended in various other formulae. Herbs which you cannot grow or pick yourself may be obtained from an herbalist or from a health food store carrying a full line of packaged herbs. (See also the source list at the back of this book.) After preparing an herb mixture to be used in a specific remedy, keep portion not immediately used in a labelled container. Allow to cool. Add 5 ounces of spirits (see recipe under "General health restoration," page 50), stir, and simmer a few minutes to form a syrup. Let this stand a day to settle, then bottle. The dose is a wineglass 3–4 times a day.

This remedy was intended for "pulmonary affections and coughs of long standing. It is admirably calculated to relieve that constricted state of the lungs which is often met with in consumption and to assist expectoration."

Variation, after Dr. W. Beach, *Scientific System of Medicine*, 1847: To the above mixture of herbs, add 2 tablespoons each of Sassafras, Marjoram and Thyme. Slowly boil a tablespoonful of the herbs (minus the Bloodroot) in 1 quart of water until it is reduced by half. Strain and add enough raw sugar or honey to make a syrup. Take a tablespoonful every hour or two.

Spasmodic asthma

Mix 2 tablespoons each of Bogbean, Hoarhound, Lobelia, Licorice root, Verbena, Comfrey root and boil in 2 1/2 quarts water, until down to 1 quart. Strain. The dose is 1 tablespoon 4 or 5 times daily.

Bronchial catarrh

Mix 2 tablespoons each of Comfrey root, Boneset, Mint (or Catnip) and Coltsfoot. Simmer 1 tablespoon of this mixture in 3 cups water until reduced by half (about 30 minutes). Strain and let cool. Add enough honey to make a syrup. Take 1 tablespoonful 3–4 times a day. Or substitute Boneset with Sage and Yarrow, or Mint with Anise and

Fennel, or Coltsfoot with Pine leaves, Mullein and Hollyhock.

Hoarseness (after Mausert)

Two teaspoonsful each of Elecampane, Sage, Marsh Mallow, Fennel (or Anise) seeds, and six of Licorice root (or Sarsaparilla) and Comfrey root.

Simmer a heaping teaspoonful in 3 cups of hot water for 30 minutes. Let cool and add enough honey to make a thin syrup.

Take 1–2 tablespoons morning, afternoon and evening.

This preparation was said to soothe the hoarseness and irritations of the vocal cords.

Cough syrup

To a quart of hot water, add 1 tablespoon each of Hoarhound, Coltsfoot and Mallow (or Hollyhock) leaves; boil for 20 minutes. Then add a heaping teaspoonful of Rosemary and the thin slices of one small Onion and simmer another 10 minutes. Keep lid on. When cool, stir and strain. Add enough sugar or honey to make syrupy. Strain into a clean bottle and label: ingredients and date of preparation. The dose is a tablespoonful, sipped slowly every 3–4 hours.

Bronchial cough

6 teaspoons Thyme	6 teaspoons Elecampane
6 teaspoons Anise	6 teaspoons Comfrey root
6 teaspoons Licorice	2 teaspoons Irish Moss

Make a mixture of the ground herbs. Slowly boil a heaping teaspoonful in 3 cups of water for 10 minutes and let it stand covered for another 10 minutes. Strain. Take 2 tablespoonsful morning, afternoon and evening. This remedy has served well in coughs and colds, conditions affecting the bronchial area, and the tickling and irritation of the throat. For best results sip slowly.

Cough, bronchial or asthmatic

To the above mixture of herbs add 3 teaspoons each of Lungwort, Sarsaparilla root and Sassafras bark and double the Irish Moss quantity.

Boil a tablespoonful in a quart of water for 15 minutes and simmer 10 more. Add a chopped Onion, stir well and keep covered for 15–20 minutes. Strain and add honey enough to syrup the liquid.

For most bronchial or whooping coughs, and quinsy (sore throat), use the roots of Comfrey if the situation is more serious, the *leaves* if less serious. Prepare an infusion by steeping a large handful of the small leaves in 2 cups of hot water. Cover with saucer for 15 minutes. Stir and strain. Simmer a tablespoonful of the roots in 2 cups of hot water for 20–30 minutes. Allow to cool and strain. Of either the infusion or decoction, take a tablespoonful every hour or two until relieved.

In cases of internal bleeding 1/2 cup of the root decoction is taken every 2 hours.

Pectoral

2 teaspoons Wild Cherry	4 teaspoons Elecam-
4 teaspoons Licorice root	pane root
4 teaspoons Anise seeds	2 teaspoons Eucalyptus
1/2 teaspoon Lobelia	leaves
4 teaspoons Lungwort	2 teaspoons Irish Moss
	4 teaspoons Comfrey root

Steep 1 1/3 teaspoons in 2 cups of boiling water, cover, simmer for 5–7 minutes and strain. Allow to cool. Sweeten with honey. Take 1/3 in the morning, afternoon, and an hour before bedtime.

This remedy was said to be useful in coughs, colds and pulmonary affections, by loosening and removing phlegm.

Comfrey lozenges

Prepare 2 cups herb infusion by steeping a heaping tablespoon of the ground root (*or* 1/2 tablespoon and 1/2 tablespoon of a mixture containing Anise seeds, Thyme and Mallow) in 2 cups of hot water for 10 minutes and then straining.

Dissolve the 3 cups raw sugar and 1 teaspoon cream of tartar in the liquid. Boil the mixture to 240°F. and add a teaspoonful of butter. Continue boiling until a temperature of 312°F. is reached (measure on candy thermometer). Remove from heat and add 1 teaspoon each of strained Lemon and Orange juices.

Pour into a buttered pan. Allow to cool and thicken; when near solid cut into 1–inch squares.

Separate the pieces, dust with powdered sugar and store in glass jars.

Compound Comfrey lozenges

The ingredients are selected from herbs that are used in homemade cough syrups: Anise, Thyme, Coltsfoot, Hoarhound, Mallow, Mullein, Boneset, Mint varieties, Wild Sarsaparilla, Sassafras bark and Comfrey root. Use 1 tablespoon each of dried Comfrey root and of a mixture of other dried ground herbs.

Simmer the herbs in 2 1/2 cups hot water for 30 minutes. Keep utensil covered. Stir and strain through cheesecloth. Over 3 cups raw brown sugar contained in a porcelain pan, pour the herb tea and dissolve well. Slowly increase the heat and bring to a boil, and continue to concentrate the mixture (about 290–295° F.) until the liquid "tests", i.e. when dropped on cold water, it hardens and snaps lightly. Then pour into a buttered cookie pan and cut into desirable size.

Comfrey-hoarhound lozenges

Over 3 cups of Hoarhound leaves and stems and a cupful of Comfrey roots (cut) pour 8 cups of hot water. Let steep, covered, for 15 minutes and strain.

To 2 quarts of infusion add 4 cups of raw sugar, 1 1/2 cups of dark Karo syrup and 1 tablespoon butter. Cook all ingredients to 300–310°F. When hard-crack stage is reached, remove the scum, pour into a flat pan, and score into squares.

"Old Fashioneds"

This basic recipe for pulled mints is easily transformed into one for a soft lozenge. Prepare a large cupful of Comfrey root infusion by gently simmering 2 heaping teaspoonfuls of the finely cut roots in 2 cups of hot water for 15 minutes, stirring well before and after the steeping. Strain and rewarm to boiling. Add 2 cups of raw brown sugar and 1/4 teaspoonful of cream of tartar. Cover and heat 3 minutes. Remove cover and cook (260°F.) without stirring until a hard ball forms on a cold area. Remove from heat and pour onto a large buttered platter or scrupulously clean cold kitchen counter. Add a few drops of Essence of Peppermint and pull and spread. Dust with confectionary sugar or corn starch.

Candied Comfrey roots

Prepare a syrup of 2 cups of raw brown sugar and 2 cups of water.

Let 4 or 5 freshly gathered and cleaned roots stand in cold water for 7–8 hours. Then cut them into *thin* crosswise sections.

Gently boil the roots in 2 cups of water, covered, for 3–4 hours. (NOTE: You may add to the boiling water containing the roots 1/3–1/2 teaspoon aromatics such as Fennel, Anise, Marjoram, Mint, Catnip or Thyme.) Strain. (Save this liquid for future use.) Place the roots in the prepared syrup, stir and let them stay there for 45–60 minutes. When the root sections are clear, remove them. Dry thoroughly on waxed paper. Store in air-tight containers.

Mix the saved liquid and syrup, add more sugar to thicken, and stir well. This syrup becomes a good remedy for mild bronchial disorders or throat irritations.

Herb lozenge

This is R. Thornton's modified method for preparing a kind of herb lozenge, as given in the *Family Herbal* (1810): "Boil some cut Comfrey roots (and other herbs) in hot water. Boil some sugar to a feather height. Add your juice to the sugar and let it boil till it is again the same height. Stir it till it begins to grow thick, then pour it on to a dish and dust it with sugar and when fairly cool, cut into squares. Excellent sweetmeat for colds and coughs."

The root enters into remedies for bronchiectasis (a condition of dilation of the bronchioles and the bronchi) and pleurisy ("dry" inflammation of the pleura without fluid formation or effusion collection of fluid between the layers of the pleura). In pleurisy, expectoration is absent or slight with stabbing pain in the side which becomes worse on breathing and moving.

Remedy for bronchiectasis

A tablespoonful each of Thyme, Elecampane, Angelica and Comfrey gently boiled 30 minutes in 3 cups of water, allowed to cool, strained and taken in tablespoonful doses, mixed in a little water, 3 or 4 times a day. Best taken in teaspoonful amounts and sipped slowly. In more serious cases (e.g. with accompanying fever) include half a tablespoon of Boneset in the formula but begin with dosage of 1/2 tablespoon and increase gradually.

Remedy for pleurisy

Combine 2 teaspoons each of Yarrow, Boneset, Mint or Catnip, Hepatica, Sarsaparilla, Licorice (optional) and Comfrey. Prepare and take as above.

Pleurisy remedy

1 teaspoon Milkweed	3 teaspoons Licorice root
3 teaspoons Elecampane root	3 teaspoons Comfrey root
2 teaspoons Boneset	
1 teaspoon Irish Moss	2 teaspoons Elder flower

Except for the Elder, all herbs are finely ground. Simmer a heaping teaspoonful of the mixture in 3 cups of hot water for 10 minutes. Allow to cool and strain. The dose is 2 tablespoons mid-morning, mid-afternoon, and in the evening.

Digestive and urinary tracts

British herbalists have long recommended Comfrey for its demulcent/astringent properties and its "binding, knitting or contracting effects" in ulcerations and hemorrhages of the kidneys and stomach as well as the lungs. For this purpose, it is often combined with astringents such as Cinquefoil, Goldthread or Wild Geranium.

Internal ulcer

1 part Periwinkle   2 parts Red Clover
3 parts Coneflower   1 part Mallow
2 parts Comfrey

Grind the herbs. Simmer 1 tablespoon herbs in a quart of water for half an hour. Let cool, stir and strain. Fill half a cup with brew, fill cup with tepid water, and sip slowly every 3 or 4 hours. Eat very little during the day and evening. Alternate the herb liquid with similarly diluted vegetable juices. Avoid all salted and spiced foods, white sugar and starches, and fatty meat.

European physicians and herbalists have prescribed and used a decoction of Comfrey root (plus aromatics and astringents) for various ulcers, alone or following the above formula, and as a detoxifying agent, or blood cleanser. The root decoction (2 tablespoons to 1 quart hot water) is taken in 4- to 6-ounce dosages 3 or 4 times a day.

I have known of persons ailing with benign gastric ulcers who have been healed with a mixture of one strong decoction of Comfrey root and leaves and Alfalfa with one part cabbage juice. They fasted two or three days a week at first, later ate sparingly, and rested as much as possible.

Earl Conroy has suggested in *Herald of Health*

that for internal ulcers one should take the young leaves in salads, juices and teas.

Blood purifier

Mix equal parts Yellow Dock root, Dandelion root, Burdock root, Yarrow, Wild Sarsaparilla and Comfrey root. Boil 2 tablespoons of the combined herbs in 3 cups water for 1 hour. Strain and take a tablespoonful 3 times a day.

The Standard Process Laboratories, of Milwaukee, Wisconsin claim their "Comfrey-Pepsin Capsules" are used by medical doctors for its "wonderful healing effect on the mucous membranes of the intestinal tract. The sticky Comfrey holds the pepsin against the mucous stomach lining to aid healing."

However, some thirty years ago, I had prepared capsules containing Comfrey, Irish Moss, Mint, pepsin, Elm Bark and charcoal, a product which several physicians (of the "old school") had prescribed for their ulcer-prone patients. (Another of my products minus the Comfrey, was called Herbazyme).* But while Comfrey and the other ingredients are of excellent service in most cases

---

*Several years later I (as Natura Products) submitted to the Federal Food and Drug Administration the following labelling for a proposed product called Carageen or Herbal Mull Powder.

Ingredients: Slippery Elm powder 8 parts, Charcoal 2 parts, Irish Moss 4 parts, Comfrey 1 part, Licorice Extract 1/2 part. An emollient and demulcent powder useful in stomach and intestinal disorders . . .

Directions: One teaspoonful mixed in water 3 times a day and at bedtime. . .

of this ailment, the ulcerite must give diet top priority. Excluded from all meals are the four whites—flour, sugar, (and jams and jellies), salt (and all salted foods,) and vinegar; fried and over-boiled foods; starchy nothing—pastries, white bread, crackers, doughnuts and white or instant rice; spices (pepper, mustard etc.) pickles, ketch-up; all soft drinks, alcoholic drinks, tea, coffee and similar "beverages"; and fatty meats.

Comfrey root, affirms W.T. Hewitt, of the English National Institute of Medical Herbalists, "is the most powerful healing agent in existence." And in his *Universal Herbal* Thomas Greene says that the dried and powdered root is "good against fluxes of the belly, attended with growing pains and bloody stools. It removes the inflammation, eases the pain, and stops the bleeding of the piles." It is also helpful in cases of rectal itching. Its mucilage is of great service in acute irritation or inflammation of the urinary passages.

Diarrhea

1. Simmer 1 tablespoon of ground Comfrey root in a quart of milk or water (or equal parts of both). Drink a wineglassful every hour.

2. Children's complaints: Mix equal amounts of Comfrey root, Sumac berries, Catnip or Mint, Hollyhock leaves and Raspberry or Strawberry leaves. Steep a teaspoonful in 1 cup hot water for 1/2 hour. Take 1–2 teaspoonsful in warm water every hour until relieved.

"Rupture of bowels"

Mix 1 tablespoon each of Comfrey root, Marsh Mallow, Wild Geranium root and Sumac berries and simmer in 2 quarts hot water for 1/2 hour. Strain. Take a small cupful 4 to 5 times daily.

"Rupture of bowels" (a modified formula)

Mix 2 tablespoons each of Comfrey root, Marsh Mallow or Hollyhock, Woundwort, Blackberry root, Wild Geranium root. Simmer in 1 1/2 quarts of hot water for 30–45 minutes. Strain, let cool and take 1/2 cupful 3 or 4 times a day.

The remedy was believed to "give entire satisfaction in all cases suffering from rupture, whatever the cause. Perseverence must be sustained, if need be, for a period of 2–3 months."

Varicocele, hernia varicosa (a modified formula)

Mix 2 tablespoons of Comfrey, Balmony, Rupture Wort and Yarrow. Simmer the mixture in 2 quarts of water for 1/2 hour. Allow to cool. Stir and strain.

To a half cupful, add an equal amount of water and take 5 times daily.

The originator of this formula states that it "has been well tried, and given great benefit insomuch as there has not been any return of the trouble unless the patient persists in his old habits."

Rectal wash for hemorrhoids (and bleeding piles)

Mix equal parts Comfrey root, Witch Hazel bark, Oak bark, Cinquefoil, Sumac berries, Blackberry root; optional are Plantain, Sweet Fern and Bayberry. Boil 1/2 cup herbs vigorously in a quart of water for 15 minutes, keeping container covered. Simmer another 15 minutes. Allow to grow tepid. Stir and allow to settle. Pour off half of the supernatant liquid into an enema bag and inject the warm liquid into the rectum. Do this 3–4 times a day. The remaining decoction should be refrigerated. When needed, remove and strain again, add hot water and inject once more.

Rectal wash #2

1/4 cup Oak bark
1 tablespoon Sage
2 tablespoons Willow Bark

2 tablespoons Life Everlasting
1/4 cup Self heal
2 tablespoons Comfrey root

Boil a level tablespoonful in 3 cups of water for 5–7 minutes. Let stand until tepid and strain. Use as rectal enema morning, afternoon and before retiring.

Kidney remedies

1. 1 part Comfrey root
   2 parts Mint (or Catnip)

   1 part Horsetail
   3 parts Corn Silk
   5 parts Dog grass

Of this mixture, steep 1 teaspoonful in a cup of

hot water for 20–30 minutes. Stir, strain and drink 4 times a day. The powdered herbs may be stuffed into #100 gelatin capsules, (see page 51). One capsule is taken 4 times a day. Follow with an aromatic tea (Mint, Marjoram or Catnip).

2. This formula had been recommended for hemorrhage of the kidneys by American medical practitioners early in this century.

Prepare a mixture of equal parts Wild Geranium root, Witch Hazel twigs, Bayberry twigs, Gentian root and Comfrey root. (At first, use only half the amount of Gentian.) Boil 2 tablespoonsful in 2 quarts of water, reducing to approximately 3 cups, i.e., less than half. Allow to cool, stir and strain. Add only enough raw sugar or honey to sweeten. Take 1/2 cupful diluted with water 4 times a day.

If by capsules, take as above indicated.

English practitioners prescribe the above in liquid form.

Wild Geranium 2 drams (tsp.)	Bayberry 2 drams (tsp.)
Bistott 2 drams (tsp.)	Comfrey 2 drams (tsp.)
Witch Hazel 2 drams (tsp.)	Tr. Gentian 4 drams

Place in a 12 ounce bottle and fill with water. Take 1 teaspoonful every hour until the hemorrhage stops.

3. Cystitis; or when urine is thick with mucous and purulent discharge; when diagnosis indicates agonizing pain in kidney area possibly because

larger-sized, roughly edged stone formations have difficulty of passage through the kidneys and bladder; or when there is constant desire to urinate but with very little urine passed accompanied by a burning sensation.

Mix 1 tablespoon each of Gravel root, Joe Pye herb, Bearberry, Parsley leaves or root, Comfrey root and Mallow or Slippery Elm. Boil the herb mixture in 5 cups of water until reduced to 3–4 cups. When cool, stir and strain. Take 1/2 cup 3–4 times a day.

A companion treatment that offers relief to the patient is covering the kidney area with a hot, wet towel and/or hot water bottle.

However, it is equally important that one should first fast for 3 to 5 days, drinking only a weak tea of Mint or Catnip; and then adhere as closely to a vegetarian diet as possible, and abstain totally from processed foods.

Connective tissue

Dr. Eric F. Powell (*Health From Herbs*) suggests the following for problems of connective tissues or cartilage (strained or damaged):

2 tablespoons Tincture Comfrey

1 tablespoon Tincture Drosera

20 drops Tincture Arnica

Directions: Take 5–6 drops in a tablespoonful of tepid water before meals 3 times a day.

Dr. Powell says: "I strongly recommend the above formula for all skeletal disorders from poor bony structures and weak bones to promoting the healing of fractures and diseases affecting the bones in general."

He has included Bittersweet (*Dulcamara*) in the remedy when "bone, cartilage, ligaments and periosteum are affected by cold, aches and pain."

General health restoration

Many old-time remedies contain herbs or roots in a base of wine. This liquor has been used for thousands of years to extract the major active ingredients (and thus, the medicinal property) which water will not. The result is a far less stimulating base than a pharmaceutical tincture prepared with true alcoholic spirits. Use an unsweetened white wine.

General wine bitters

1 tablespoon each Comfrey root, Solomon Seal root and Spikenard root and 1 teaspoon each Gentian root, Burdock root and Chamomile flowers are mixed. Cover the bruised roots and Chamomile with boiling water and let stand covered in a non-aluminum container for 2–4 hours. Add 1 quart of sherry wine and let macerate for 2 weeks. Express and strain.

One or 2 tablespoons may be taken in a little water 4 times a day for loss of appetite, "that run-down feeling," and after prolonged illness. For leucorrhea, 1/2 to 2 ounces may be taken 4 times a day.

Restorative wine bitters

Mix 1 tablespoon each of Comfrey root, Sarsaparilla root, Gentian root, Yellow Dock root, Burdock root and Dandelion root, and 1 teaspoon each of Cardamon seeds, Fennel seeds and Chamomile flowers. Prepare as above. This is an invigorating tonic and useful in female disorder (leucorrhea, menopause). To use as a pectoral add enough sugar or honey to make a syrup.

Take a half wineglassful (4 ounces) 3–4 times a day.

Herbal powders

Another way of taking herbs: powders. I have often advised the members of my herb-study classes to finely powder their thoroughly dried herbs* whenever possible in a coffee/spice grinder; sift and stuff the product into druggist's empty #100 gelatin capsules.

Leaves powder easily, but not all roots do. Powdered roots may be obtained from an herbalist or herb-selling health food store.

Separate the parts of the capsules and stuff as much of the powder into the larger part as will be received. Insert a little powder into the small part and push both parts together to fit snugly.

Another way of taking the powder: mix the powder in water, with enough honey or sugar to sweeten and take by the teaspoonful as directed

*Powdered herbs serve also as salt substitutes and food seasoners, ingredients of homemade bread and pastry, and as food supplement.

(i.e., morning and night, or 3 or 4 times daily).

NOTE: The powder may represent one herb or a combination but should be measured in approximate doses.

You may also prepare 10 or 20 "pulvules" (as we called powders prepared for customer patients in my early pharmacy practice). One merely encloses required amounts of the powder, previously sweetened with powdered sugar or milk sugar, within 4 x 5 inch white powder papers. The edges are folded over and sealed, to prevent the powder from leaking.

INTERNAL VETERINARY MEDICINE

In recent years Comfrey has made its mark in British veterinary circles. It has been used as a preventative and treatment for scour (diarrhea or dysentery), wrote Lawrence D. Hills, in *Health from Herbs*, "in mares and foals of a famous racing steed. A racehorse is liable to digestive upsets in the same way as human beings." Certainly horse breeders recognize "summer diarrhea" as a serious and ever-present problem.

Mr. Hills tells of a Mr. E. V. Stephenson who fed his horses Comfrey leaves as a source of high protein (3.4 percent), low fiber (1.5 percent), and calcium (1.41 percent of the dried material) for good bone structure. His race horses, he says, thrive on this herb. Moreover, Mr. Stephenson discovered that since only that farm which fed Comfrey to its animals "remained entirely free from scour and allied digestive troubles," he expanded the herbs usage to his entire half million dollars' worth of stallions and mares.

Pig breeders have noted that the leaves are a most suitable combination of food and remedy for young pigs who are changing over from sow's milk to meal, and for "those affected by the *bacillus coli* when three days old or three weeks after weaning." The herb is equally effective with adult pigs, especially for digestive disturbances, after-effects from a change of diet, and for post-farrowing scour.

EXTERNAL USES

Most herbalists have heeded the words of the renowned, unparalleled herbalist Culpepper, especially where Comfrey is concerned. The following recommendations which his *Theatricum Botanicum* offers are employed by herbalists throughout the world. He advised using "the distilled water for the same purposes also for outward wounds or sores in the fleshy or sinewy parts of the body and to abate the fits of agues [pain] and to allay the sharpness of humors." A decoction of the leaves, he said, is "not so effectual as the roots.

"The roots being outwardly applied, cure fresh wounds or cuts immediately, being bruised and laid thereto; it is especially good for ruptures and broken bones so powerful to consolidate [a synonym for Comfrey is *consolida*] and knit together that if they be boiled with disservered pieces of flesh in a pot, it will join them together again."

The roots "taken fresh, beaten small and spread upon leather and laid upon any place troubled with the gout presently gives ease ... [also to] pained joints and tends to heal running ulcers, gangrenes, mortifications for which it hath by often experience been found helpful."

Wounds, ulcers, bruises

Both the root and the leaf enter into several applications. As an application to bruises, fresh wounds and severe cuts and ulcers, the peeled fresh root may be pounded or bruised, dipped into warm water for 2 or 3 minutes and applied to the affected area. Most external ulcers will respond satisfactorily if a *fresh* macerated leaf is applied to them 3 times a day. Be sure to bruise the leaf's veins which will come into direct contact with the affected part, to better heal it. Leaves are similarly employed topically by British veterinarians to open wounds and stubborn sores and ulcers.

The bruised wet leaf is also a safe and effective styptic—it stops all kinds of minor bleedings.

A poultice may be prepared by blending several large leaves, previously washed clean. The resulting okra-like, mucilaginous mass is placed on several thicknesses of cloth (white shirt material will do) and applied to the area. No juicer? Macerate the leaves on a handgrater, or use a vegetable or meat grinder. Or place several layers on a board and beat with a hammer.

A hot fomentation of the leaves reduces the

pain and swelling of bruises and sprains; cooled, it is also good to heal burns and fresh wounds. (The leaves may be substituted for Mallow or Hollyhock when they are indicated in external ulcers.)

When needed for internal hemorrhoids, the root may be boiled with Sumac berries, Wild Geranium and Cinquefoil roots (all equal parts) and the warm, strained decoction enemated 3–4 times a day. The liquid may also be applied to all skin affections and bruises.

English herbalists have used the leaves with good results as a warm poultice in periotitis, inflammation of the external covering of the internal ear's temporal bone. A hot fomentation quickly promotes suppuration of boils and abscesses, and soothes the pain in the tender inflamed (suppurating) areas. Tepid, the cataplasm may be applied to raw, indolent ulcers.

Even employed as a jell, the powdered root "dissolved in water to [form] a mucilage, is far from contemptible for bleedings and fractures, whilst it hastens the callus of bones under repair."

Irritations

In general skin irritations, damaged skin or insect bites, wrap a bruised leaf, quickly dipped in warm water, around the hand or leg. Keep bandaged 3–4 hours and repeat with *fresh* leaf. (For winter use, store several plastic-wrapped leaves in the vegetable section of the refrigerator.) In the case of slowly healing sore or ulcer, continue the treatment for a week or two even after healing is accomplished.

An ointment may be prepared by slowly simmering a tablespoonful of the dried root in a cup of *unsalted* lard, lanolin or suet for 10 minutes and straining. Allow to cool before refrigerating. A few drops of tincture of Benzoin Compound may be incorporated into the cooled ointment.

A plain ointment, states the celebrated Swiss Dr. Vogel, prepared by mixing the root powder in a cosmetic base, will help to overcome "wrinkles" or "crow's feet."

A thick paste may be prepared by mixing well the powdered root with Castor Oil. This preparation is soothing to rough or chafed skin. A thinner paste may be obtained by less powder in a lighter oil like Sesame or Safflower.

Once thoroughly dried, the powdered leaf or root alone becomes an excellent powder for open wounds, cuts, or athlete's foot.

—Or use equal parts of cornstarch and the powdered root.

—Or 1/4 part each of plain talc and of zinc oxide, and 1/2 of the powdered root.

—80 percent Comfrey powder and 20 percent zinc undecylenate (obtainable at your pharmacy).

Athlete's foot powder: Mix together 1/8 teaspoonful *each* of benzoic acid, salicylic acid, precipitated sulfur and thymol iodide. Then incorporate with 1/2 cup powdered Comfrey root. Use the powder only for athlete's foot morning and night. Dust between the toes. Do not use near the eyes. (May be used for ringworm also.)

Athlete's foot ointment: Mix together two teaspoons of Comfrey root powder, 1/2 of salicylic acid, 1/4 each of benzoic acid, precipitated sulfur and thymol iodide and incorporate well into 2 ounces each of *yellow* vaseline and lanolin. Rub into the area morning and night. Use this ointment for dry and wet itching eczema.

CHAPTER 4

COMFREY AS FOOD

A source of food

In the past twenty years, the leaves of this newly discovered old-fashioned plant have come into vogue in this country as an acceptable food plant. For centuries the country folk of European lands had partaken much of this well-known almost weedlike, profuse grower. They needed no ballyhoo of well-meaning writers in health-oriented magazines or authors of health-diet books to tell them that Comfrey's leaves were to be eaten or used as a simple remedy. As so often is the case, these farm people may have witnessed their animals relishing this green fodder and later displaying no ill effects; then, I presume that they were as adventurous as my herbalist grandfather, thinking (in his words), "If the herb is good for the beasts then it must be good for us." Good examples of such philosophy are Alfalfa, several Clovers, green grass, Nettles, Amaranth ("pigweed"), Lambsquarters, and others.

These edibles demonstrate that there are several dozen highly nutritious plants growing in Mother Nature's immense garden free for the asking. The long list includes the exemplary Dandelion, Burdock, Yellow Dock, Milkweed and Plantain. If you find their taste unusually sharp or somewhat bitterish, as some have found Comfrey, I remind you *that bitter to the taste is sweet to the stomach* and vice versa. Indeed the bitter taste usually results from a high percentage of blood-fortifying minerals—calcium, iron and potassium.

If certain newcomers to the realm of eating herbs, or more specifically the leaves of Comfrey, will tolerate this unfamiliar experience, they, too, will gradually overcome a refined taste long influenced by improper dietary and seasoning habits, e.g. overcooked foods and the subsequent need for salt, spices, condiments, vinegars and the like. In time, the anti-herb eating bias slowly disappears, especially after one fares on the more "respectable" Borage whose leaves, as rough, hairy, and mucilaginous as Comfrey, are also eaten as a "Spinach substitute" and relished. Many weed-eaters have found a variety of acceptable ways, especially in these inflationary times, to guard and stretch the food dollar while preserving one's health.*

It's important that one consider young early leaves, about four to five inches long, for purposes of food or herb tea. (There are a few cases, of course, where the larger leaves are used for internal purposes, as we note under Comfrey's medicinal properties and the various remedies). In fact, many find the large, fully grown leaves too coarse; to the uninitiated they seem rather unpalatable. The taste of the Spring leaves may be "pronounced" or "distinctive" and some writers do advise cooking them in two waters—a waste of health-preserving nutrients and precious time— and "flavoring" them with Chives and salt. (Chives are usually eaten uncooked; salt is a dangerous/poisonous irritant to the body, greatly decreases normal digestion of food and now is forbidden to all who suffer from obesity, kidney and heart troubles, high blood pressure, etc.)

Comfrey in cooking

There are several ways to let Comfrey leaves serve you as a worthwhile food source. The *uncooked* leaves may be included whole or sliced thin in a mixed vegetable salad. In the beginning you may want to modulate the taste of the leaves by first soaking them for a few minutes in an herb vinegar of your choice (Basil, Oregano or Garlic, or a mixture of all three). Shake off the

*For further suggestions, see my *Eat the Weeds,* Keats Publishing Co., New Canaan, Ct, 1974.

excess liquid. Cut the leaves in 1-inch squares, and add them to the salad. In the future cut them in thin slices or leave whole.

They may be steamed alone or with other vegetables in a Double boiler in as little water as possible. Be sure to use the remaining liquor as a drink between meals, a soup or stew ingredient, a liquid in bread-making or reheating other vegetables. Heat only long enough to soften the tissue—the less heating time the better. Extra seasoning can be provided with a judicious pinch of Basil, Mint, Marjoram or Savory.

The leaves should definitely not be boiled, or cooked too long. They are not to be considered as a "Spinach substitute" or equally disreputable potherb. Both cases require drowning the herb in water and boiling—spoiling them, with the inevitable destruction and loss of vitamins and minerals. Indeed boiled or long-cooked Spinach, French Sorrel and Rhubarb rank high in my list of No-No foods. They should be steamed but a minute or two—and even eaten uncooked!—for if cooked to an untimely death, their oxalic acid compounds become highly concentrated and are often the cause of bigger and better kidney stones.

Young Comfrey sprouts are eaten as a rough but acceptable substitute for Asparagus. During the winter months, some enterprising Comfreyites have had a continual supply of the sprouts by blanching them, i.e., forcing their growth through loose rich soil in the cellar. But I prefer the lacy leaves of Parsley or Carrot that comfortably sprout on my sunny window sills from October to May: no blanching, no fuss. And how about using the poor man's Asparagus—Milkweed shoots—that abound in New England and south to North Carolina, and the flowers and the later pods?

Over the years many Comfrey devotees have incorporated the finely ground leaves—their choice is the small-sized ones—in soups, stews, casseroles, omelettes, scrambled eggs, chicken and fish croquettes, hamburger and even sandwiches. When taken with soups or stews, the leaves, now cut into small pieces or diced or slivered, should be stirred into the cooking food long enough to be softened adequately—and just before the soup or stew is served.

At Eastertime English country folk have often used the leaves (possibly the older ones) to flavor satisfactorily various pastries. The freshly gathered leaves were chopped fine and added to the cake mixture. Many springtime preparations— pastries as well as cold dishes, the fish and meat items—required a particular sharpness of taste. This service was rendered equally by such well-known savorizers as the garden-grown Sage and Rosemary and especially Tansy, a commonly recognized wilding. For good reason, then, that cakes made in the Spring were called Tansies. Tansy and other aromatics were then included in foods as substitutes for the costly Cinnamon and Nutmeg to "carry off bad humours" (catarrhal deposits, phlegm, etc.), to counteract the saltiness of fish and the lethargy which "the moist and cold constitution of winter has made on people."

"The (Comfrey) leaves are frequently employed," wrote Thomas Green in his *Universal Herbal* (1823), "to give a grateful flavor . . . to panada." This is a paste preparation of flour or bread crumbs in water or broth, which is intended as a base for meat sauces and as a binder for stuffing (forcemeat).

"Comfrey fritters are delicious. Dip the young leaves in cold water, shake them, plunge them into batter and then into the sizzling fat. The tender young leaves make a pleasant green vegetable and, chopped fine, they give a distinctive flavour if added to cakes and puddings." So says Timothy Green, in *Health From Herbs*. Other fritters may include Rose petals, Elder flowers and Mint leaves.

Assorted items: Japanese food processors use dried Comfrey leaves in an assortment of food products and "cannot obtain sufficient for their needs."

• Some herb-eaters have even sprinkled the dried powder onto their breakfast cereal.

• The European poor prize young Comfrey roots as a nutritious food and cook them.

• A combination of Dandelion, Chickory and Comfrey roots is used to prepare a much popularized "vegetable coffee" that is said to taste "practically the same as ordinary coffee with none of its injurious effects."

• Comfrey leaves have been included in popular "potassium broths" and "green drinks" which

have been highly touted as near cure-alls, as
panaceas for almost all ills known to man. (They
certainly are not.) They may be duplicated to a
lesser degree by an herb tea and far more sub-
stantially by the final products of a blender or
juicer. Naturally, potassium (and several other
minerals) are found in Comfrey but it occurs
principally combined with calcium in all foods—
nuts, salad greens, sweet and acid fruits, root
vegetables, and in the common weed-like herbs.

Lawrence D. Hills supplies these recipes by
English herbalists.

1. Comfrey flour-mixture: Mix equal parts of Comfrey, Soy and other meal with a vegetable oil
   to make a smooth paste. Heat gently for 10
   minutes, stirring. Store in a glass jar for one
   week only. This serves as a base for gravies,
   soup, and stews. Use also in scrambled eggs.
   Season with tomato sauce, herbs or curry
   powder.
2. Comfrey Root Marmalade: "Grate equal a-
   mounts of washed (recently gathered) roots
   and lemon or orange peels and leave them in
   a covered dish with an equal amount of sugar
   for a few days. Then cook with grated apple
   and sugar to a standard marmalade recipe."
3. Comfrey Au Gratin: "A layer of cooked rice on
   bottom of glass dish, add a layer of cooked
   Comfrey leaves with some grated cheese and
   a dab of butter, then more Comfrey and a
   final layer of cheese. Bake half an hour."

4. Candied Comfrey Root: "Wash the roots and boil gently until they are tender, drain and then soak for an hour in a syrup made of 1 lb. of sugar and a half of the water the roots were boiled in, and a little lemon juice. Then boil over low heat till the syrup candies. Drain off the syrup and dry the roots in a warm place."

Blendered and juiced

Health enthusiasts should consider foods, our fruits and vegetables, nuts and grains and their uncultivated, weedlike counterparts, as a source of vitamins/minerals, of added nutriment. The blender (or food processor) and juicer are excellent means to yield food supplements on demand.

Young Comfrey leaves may be blendered alone but should, at least at the outset, be complemented with such freshly gathered commoners as Dandelion, Nettles, Lambsquarter, Watercress, Purslane ("Pussley"), Peppergrass, Amaranth, Green Grass, Alfalfa, Evening Primrose (early rosettes) and the stems of Japanese Knotweed and Burdock. The beginner should first try a blend of Comfrey, Watercress and fresh mint (or Basil), plus any of these vegetable companions: celery, cabbage, carrots, Parsley, Romaine lettuce and the green tops of beets and turnips.

Blending the herb-vegetable items requires a quick brush-wash with cold water, a rinse and an immediate insert into the blender. Do not take the end-product as is. At first, to better get accustomed to a strange delight, dilute a small amount with 2 or 3 parts of water, *not* with fruit juice, as some have suggested. (Fruits and vegetables—or herbs—make a very poor food combination.) Season with powdered aromatic herbs of your choice or with an herbal salt substitute (See *Eat the Weeds*, p. 30). Stir thoroughly and either slowly sip very small amounts or take by the teaspoonful. Gradually lessen the water dilution and in 3 or 4 weeks, take the blend without water.

The remaining mixture may be refrigerated only 3 or 4 hours. Blendered herbs (and vegetables) do not keep well. Therefore prepare a liquid blend of Comfrey and its companions only as needed. Should the fibrous mixture of Comfrey not be to your liking, you may strain it and, proceed as indicated below.

The requirements for juicing Comfrey leaves are as few as with blending (blenderizing). Use the early leaves of Comfrey either alone or, as indicated, with the same recently gathered, fresh (i.e., undried) herbs—but *less* of the aromatics— and fresh vegetables. (Comfrey, too, is a vegetable.) It is best to use a juicer guaranteed not to rust.

Wash the Comfrey and other ingredients well, rinse, place them in a shallow dish, and barely cover with cold water. After a 2 minutes' soak, process them in the juicer. Insert the leaves first, then the stems, and roots or tubers last. Yes, you may and should juice carrots and potatoes with Comfrey leaves. After such insertion, add a little of the soak water.

While the resultant drink contains the genuine vitamin-mineral-enzyme extractives, that wonderfully health-protecting, chlorophylated essence of these foods, still one should not depend wholly on juices or blendered mixtures as one's sole food intake nor as a substitute for a meal. It could be considered soul food, however. Remember that herb/vegetable juices and blends are but temporary measures. By far most foods and weedy herbs per se are to be eaten whole. And the juice of Comfrey or of other foods or other liquids may be taken *before* a meal, not during or immediately following a meal.

In the beginning it is best to dilute the juice with an equal amount of water. Above all, be sure to sip only a *little* bit; swish it around in your mouth so that it mixes with salivary enzyme and thus becomes predigested. Taken this way, the juice is assimilated in the bloodstream in less than an hour. All liquefied potions are acceptable weapons during a detoxifying or weight-reducing program.

Comfrey wine

My grandfather, that Thoreauvian herbalist, used native herbal material to prepare his wines. For the most part, the ingredients consisted of the basic "honesties" like Yarrow, Dandelion, Burdock and the Raspberries and Blackberries; the discards of the vegetable patch—the darkened greens of turnips and beets, carrot tops and leaves, some potatoes and especially those oversize Borage leaves; and the highly desired all-purpose native aromatics, Sassafras, Mints, Bee Balm, Catnip, Wintergreen and Wild Thyme. He had no "still room" as did our more fortunate well-to-do neighbors, but the many shelves that bordered three sides of his cold-storage stone hut (his outdoor ice-box) housed his home-made products—the preserved foods, his medicinal cordials and elixirs, innumerable quantities of his medicinal, "God-healing" remedies, and, it always seemed to his grandchildren, a never-ending supply of wines. Odd-shaped bottles were usually discards or bartered for with the rags-bottles man—who could afford *new* bottles? And who cared? Just to catch a quick glimpse of colors ranging from purple to blood red to yellow to white fulfilled a great (and imagined) adventure. I was privileged on Sabbath eve to sip a bit of a "sweet" wine to climax my saying the bread and wine blessings. Permissible to the Harris children were Elderberry, Blackberry, Raspberry and other fruit wines unless they were spiced and mulled, for then they were drunk only during the cold winter months.

Admittedly Grandfather knew nothing of friend Comfrey; certainly he cultivated and used much of its cousin Borage, "the dark green ears with the blue flowers." When Parsley was not readily available, this herb served as a food-stuff—in salads, as a steamed vegetable, as a pot-herb in meat soup or stew and part of our Shallot-Onion relish, and as a medicinal remedy for kidney and

bronchial disorders. Aunt Esther told me that
Grandfather steeped Borage leaves in aromatic
wines for an hour or so not so much to make us
"glad and merry"; but how else overcome the
roughness of the leaves' prickly hairs? That's
probably how the Borage herb became an ingredient of his "leaf" wines. In fact, except for Comfrey's possession of allantoin, the powerful healing
agent, both herbal relatives are somewhat inter-
changeable. Thus a wine recipe requiring Comfrey leaves or roots may use those of Borage.

Comfrey bitters (beer)

Take 1 pound each of *fresh* roots of Comfrey
and Burdock, and 1/2 pound each of Wild (or
Garden) Carrot and Wild Sarsaparilla. Scrub-wash
with cold water, slice lengthwise in quarters and
cut in halves. In 3 quarts of boiling water contained in a non-aluminum container, boil the
roots down to 1 quart. Keep the vessel covered.
Strain, allow to cool down but not completely,
and add 1 cup of honey or molasses, or 1/2 cup of
raw brown sugar, and enough yeast to initiate
the proceedings. Let the mixture ferment for 3 or
4 days. (One may drink a little of the clear liquid
even after the first day of fermentation.)

A further *better bitter beer* is presented with
the addition of, or substitution of the above ingredients with appropriate amounts of fresh Dandelion or Chickory roots, or Barberry fruits or stem
cuttings.

Comfrey root wine

An unusually pleasant tasting drink or beverage for those who are well and supposedly beneficial for others who are recovering from "debilitating diseases." (One author has even gone so far as to recommend Comfrey wine "for sufferers from internal ulcers.")

You will need 4 pounds of fresh early roots, cut up before they are weighed, and about 3 pounds of raw sugar to each gallon.

The roots are washed clean, peeled, sliced and cut into 4- to 5-inch long pieces. Boil them gently until fully soft. Remove the lid and allow the strong odor to escape. Stir the mixture. Strain the decoction through muslin. Measure the liquid and return it to the pot. Dissolve 2 1/2—3 pounds of sugar to each gallon and stir frequently.

Boil the syrup for 45 minutes, still stirring, and pour into a suitable container, preferably earthenware. *Allow to cool* and when gently *tepid*, place a little yeast on toast and then onto the contents. Cover with a cloth and let it so remain no less than 10 days. From the second day on, be sure to stir it each day. Then remove it to a wine-barrel, stone jar or other suitable vessel until fermentation is completed. This wine is fit to bottle and drink in 4 to 6 months.

A variation: Substitute 1 pound of fresh Burdock roots for 1 of Comfrey. Proceed with the boiling-to-tenderness measure but when you remove the lid as above indicated, stir in 3 tablespoons of Chamomile flowers and 2 tablespoons of Sassafras bark and then allow mixture to cool.

Comfrey tinctured wine

Into a tightly covered vessel (a wide mouth bottle with stopper) containing a quart of the cheapest wine, admit 1 pound of the prepared roots as above indicated. Add a teaspoonful each of coarsely ground aromatics like Sage, Mint, Thyme, Fennel, Anise and other seeds, and/or Marjoram, and, if taste demands, of the stronger spices (Cloves, Cinnamon, Ginger, Mace). Keep stoppered, and let stand for 2 weeks, shaking it twice a day.

This procedure is after John Partridge's *Treasury of Commodius Conceits and Hidden Secrets*, 1586. The resultant liquor was to be "stilled" (distilled) via an alembic distillation apparatus. "Of these waters," Mr. Partridge wrote, "the virtues be these—it comforteth the spirites and preserveth one greatly ... make one seeme young very long." The dose of the distilled wine was "one teaspoonful of this water fasting."

Compound wine of Comfrey

The "Restorative Wine Bitters" of the last century is now called Compound Wine of Comfrey; it was listed in *The Household Physician* by Dr. Ira Warren in 1860. The formula was often prepared by herbalists (even in my drugstore apprentice days) and prescribed by medical physicians for leucorrhea, vaginal disorders, and other female complaints.

Mix 2 tablespoons each of roots of Comfrey, Solomon Seal and Spikenard (or Sarsaparilla); 1 tablespoon each of Colombo, Chamomile flowers and Gentian. The herbs were covered with boiling water and allowed to stand in a covered vessel. Two quarts of Sherry wine were added and allowed to macerate 14 days. This was then expressed and strained through cloth.

The dose was 1/2 to 2 ounces taken 3 or 4 times a day.

A contemporary of Dr. Warren, a Dr. W. Beach, had written in his *The American Practice* (1847) that this same Wine Bitters was "a useful tonic in all cases of debility, particularly that peculiar to females." It is pectoral and corroborant (an invigorating medicine or tonic.)

Another Compound Wine of Comfrey or "Restorative Wine Bitters" was prepared and sold in 1877 by the Henry Thayer Co. of Cambridgeport, Mass. The ingredients: 2 tablespoons each of Solomon Seal root and Unicorn Root and fluid extract of Comfrey, and 1 tablespoon each Chamomile flowers, Colombo, Cardamom seeds and Sassafras, 4 oz. of Alcohol and 4 pints of Sherry wine. The uses and doses as above mentioned.

Of course wine-making was a common practice in all British and early Colonial households and the final wine often became the base for an aromatic, spiced medicinal wine called "hypocras" (hippocras). This was a non-distilled item and so prepared. Into a gallon of Claret wine went 2 ounces of mixed spices—Cinnamon, Ginger, Nutmeg and Cloves, one or two small cut oranges, and a large sprig of Rosemary. The herb-spice items were ground into small pieces and put into the wine, the container then "close stopped" for 3 to 5 days. The mixture was shaken and strained through cotton, linen or flannel, yielding a highly desired cordial drink comparable to the Swedish *glog*.

Another *hypocras* formula: To a quart of white wine add 1 tablespoon Cinnamon, several leaves of dried Marjoram, and 1 or 2 slices of Lemon. Let this stand a day or two, shaking it often and strain.

Herb teas
(also called tisane, tizane, or ptisane)

These are basically infusions of various, that is, single herbs, or mixtures of herbs. They are considered a healthful substitute for the astringent tannin- and caffeine-containing pekoe tea. The herb tea may be drunk any time of the day, but never during meal-time or directly after the meal (lest the gastric juices be diluted and the food not afforded optimum digestion).

My perennial statement to likely herb tea drinkers: A warm herb tea keeps one cool in summertime and more comfortable in winter. The infusion quenches the thirst most admirably. Indeed when I undertake my monthly three-day fast, diluted herb teas are my only food intake.

An herb tea should be taken warm to tepid; iced, it is therapeutically inert and therefore undesirable.

The herbs:

1. Non-aromatics, including the leaves of such (dried) *native* herbs as Linden, Boneset, Goldenrod, Stinging Nettles, Alfalfa, Clover species, Sweet Birch, Yarrow, Rose petals and leaves, Sassafras, Sweet Fern, Sumac flowers and early fruits, Meadow Sweet and of course *early* Comfrey.
2. Aromatics: Assorted Mints, Bayberry, Chamomile, Catnip; food seasoners (Thyme, Savory, Lemon Balm, Marjoram, Basil and others of your choice); and the dried rinds of Orange, Lemon, Tangerine and Lime.

Take equal parts of two or three dried herbs in each category, and coarsely grind them up. Mix them equally; a teaspoonful of the mixture should clearly indicate participation of each herb. Label each container of herb mixture with the ingredients and the date. Note: for best results, use the dried, not the fresh, herb.

To prepare an herb tea: Bring to a boil the required amount of fresh tap water—or even rain water. Place a teaspoonful of the herb or herb mixture into a cup, slowly add the hot water, stir the herbs about 25–35 times, and cover with a saucer. In 8 to 10 minutes stir and strain. If the liquid is too hot, stir again and leave unsaucered a few minutes. Sip a little *slowly*, about 1 or 2 teaspoonsful at a time, and swish in the mouth before swallowing.

Best not to add milk (as some of my English friends do) or sugar or honey. Honey may be used at the beginning of this "experience" but should be abandoned as early as possible.

Unless the resultant liquid is prescribed for strictly therapeutic purposes, do not use large leaves nor boil them. It is best not to refrigerate any excess. Prepare each cupful fresh although you may take a thermosful with you. For the usual herb tea, always use ground dried leaves, not the root. But if taken for gastric (stomach) ulcer or lung ailment, a cupful of the steeped roots (1 or 2 teaspoonsful in a cup of hot water) may be taken several times a day. (Further information on the range of herbal drinks will be found in *The Herb Tea Book* by Dorothy Hall, Keats Publishing Co., New Canaan, CT, 1981.)

CHAPTER 5

CULTIVATIONS, COLLECTIONS AND PRESERVATION

The varieties of Comfrey

Common Comfrey, *Symphytum officinalis*, has an average height of 3 feet. It blossoms from June to September; flowers are white, blue-purple or rose. A native of Europe and Asia. The *argentum* variety produces white variegated leaves.

Prickley Comfrey, *S. Asperum*, 2 1/2 to 5 feet. Its stems display coarse prickles and the leaves are covered with stiff, prickly hairs on both sides. The flowers at first appear rose, then blue or pale purple and are smaller than those of above. Russia to Iran.

S. var. rubium. English catalogues list it as a garden variety, for cultivation in loamy soil under trees. It produces deep red flowers.

Tuberous Comfrey, *S. tuberosum*. It is common

in England, Ireland and Wales, and produces yellowish-white flowers and a stout rhizome.

Caucasian Comfrey, *S. caucasicum*, is a more ornamental species than the Common Comfrey. Even more of the numerous gentian-A blue (and some pinked-blue) flowers will appear (from June to early August), if the dead flowers are removed.

S. Peregrinum is also a more showy plant than the Common Comfrey, with pendulous flowers ranging in color from rose to blue. The variegated form presents leaves with creamy white markings.

Like several other members of its family, the *Boragiaceae*, Comfrey is characterized chiefly by the white harshly stiff hairs that cover its stems and leaves. Dr. O. Phelps Brown warns us to beware these prickles for to "touch any tender part of the hand, face, or body will cause it to itch."

The large alternate leaves are darkish green and resemble something between a foxglove and horseradish. The drooping, bell-shaped flowers of various species range from white to reddish-purple to blue and appear in clusters on short stalks, usually all turned towards the same side. The seeds number four, are dark brown, shining nutlets.

The long fleshy root of a mature plant is usually about one inch thick, branched, white inside and blackish outside. (Hence the synonym blackwort).

THE FAMILY INCLUDES:

Borage, *Borago officinalis*, L. The fresh large leaves are usually steamed and eaten; the dried leaves and flowers have been used as a demulcent and diaphoretic in cough remedies.

Lungwort, *Pulmonaria officinalis*, L. This herb has great repute as an effective soothing expectorant in bronchial problems.

Hounds Tongue, *Cynoglossum officinale*, L. This strongly scented biennial contains two poisonous substances *cynoglossine* and *consolidin*, and when used as a sedative-demulcent, should be used with caution. Caution: The older, larger leaves are often mistaken for Comfrey's and if eaten fresh, there may result harmful after-effects of vomiting, stupor and/or sleepiness.

Viper's Bugloss, *Echium Vulgare*, L. This rough and robust relative, although less beautiful than its more stately cousins, offers its demulcent properties in bronchial and urinary situations.

Alkanet, *Alkanna tinctoria*. The roots yield red coloring substances.

Forget-me-not, *Myosotis palustris*, L. Its leaves are most useful in pulmonary affections.

Gromwell, *Lithospermum officinale*, L. The seeds had formerly been used as a stimulant to the urinary organs. Strong infusions of various species became the Shoshone woman's contraceptive.

Garden Heliotrope, *Heliotropium arborescens*, an unusually sweet-scented plant, has been used not only as a source of a most fragrant oil which is often used as an ingredient of sachets and "sweet pillows," in perfumery and cosmetics, but also as a parasiticide. DANGER: Heliotropin, the active constituent of this plant, exerts a depressant effect on the cerebro-spinal axis.

The foremost external feature of this coarse homely herb—the sharp, bristle-like hairs—is characteristic of its cousins and is sharply noted also in Bugloss, Borage and Hounds Tongue. Comfrey's identifying feature, the signaturing stiff hairs indicate to the observant herbalist that this herb should be used to soothe varied irritated conditions. Like heals like. This book has included several remedies that show Comfrey's application as a vulnerary and pain-relieving soak for hurting and inflamed situations.

Similarly the heavy down of the soft hairs of some herbs like Marsh Mallow, Hoarhound and Mullein provide, according to a diagnostic system known as "The Doctrine of Signatures," clues as to their healing properties. The system called the Doctrine of Signatures has been my means of teaching the folklore and many uses of herbs for over thirty-five years. Signatures may be placed in several categories and usually indicate the healing properties of a given herb. A habitat of swamps, or wet lowlands indicated plants growing there as remedies for feverish colds or rheumatic genito-urinary disorders, while plants found in mucky soil were to be used to remove mucous secretions and those in sand or gravelly soil, to help dissolve and expel gravel and stone. Yellow-colored flowers often indicated a plant's use in jaundice, liver and gall bladder problems. The gummy exudations of evergreen trees have long been indicated in cough remedies and remedies for external sores.

Observant gardening members of my herb-study classes have suggested still another signature for Comfrey. This herb grows in profusion and therefore should be employed in profusion just as we use Dandelion, Lambsquarter and Yellow Dock and other ubiquitous plants almost every day.

The flannel texture of these herbs proclaims their softening effect upon a diseased area, internally and externally. For that reason they—and Woundwort and Hollyhock—were employed in Revolutionary times as a lint substitute for dressing cuts and wounds. And one or more of the aforementioned herbs are to be included in most internal herbal remedies for their emollient—tissue-softening—effect.

Comfrey's invaluable effectiveness is pinpointed by its dominant external feature: the stout coarse hairs. They are indicated in painful situations, i.e., the herb is generally employed to ease irritations. True, the herb is listed as an "anodyne" or pain-reliever, but it is more important that Comfrey serves as a choice vehicle that soon overcomes the cause of the pain. (The hairiness of herbs like Nettles, Sumac and Sundew signify their application in a variety of irritating or painful internal disorders—e.g., "stitch in the side." Hops, too, have long been employed for their calmative and anodyne properties.)

Herb users should carefully note two other of Comfrey's specific signatures—the glutinous, mucilage-rich substances which the leaves produce when chewed, and the hollow stalks. Thus both characteristics indicate the herb's Mallow-like* emollient action as a most worthy demulcent and expectorant in cough remedies and as an especial means of removing catarrhal deposits from the hollow bronchial tubes and the intestinal tract.

Comfrey's cultivation:

One or two plants will suffice an average family's needs. Indeed a single planting of a fresh root will provide you with enough material for all your neighbors and friends "and your sisters and your cousins and your aunts." But this prolific grower can become a pest and if not attended to may take over far more profitable garden space than it deserves. Therefore, select the proper place for the herb in your garden, preferably in the background of other tall perennials, or near the Hollyhocks. If it is to serve as an "ornamental" plant in the garden and is allowed to spread, it will be an almost impossible task to eradicate it. Comfrey has that never-say-die perseverance when once it is established and if left untended, new plants will arise from the seeds or severed roots portions. Members of my herb-study class have

*Mallow is derived from a Greek word, *Malake*, "to soften, to heal." Hence its use in inflammation and irritations of the bronchial, intestinal and urinary passages.

cultivated Comfrey via root cuttings in waste places full of composting material and along brooksides.

Comfrey prospers under most conditions and grows in all but the coldest parts of our country. The plant would like full sunshine but will do fairly well in partial shade between trees in a grove. It will adapt quite easily and multiply even in partially shaded *moist* woodland and rear-garden areas. Soil requirements range from any good/rich type to limestone or clayey loam or sandy loams with a pH at least six or higher. If the soil is clayey, mulch with peat. Spring and fall plantings are equally successful.

It is possible to obtain good growth even in poor or uncultivated (garden type) soil. The far-reaching and almost everlasting root system will stretch into the subsoil and supply the plant with water and nutrients.

Propagation is an easy matter and more often than not takes place by the division of the roots (my method). Comfrey enthusiasts report sowing the seeds in spring, about 2 to 2 1/2 feet apart, covering them with a thin 50 percent layer of composted material and soil, and watering well. But each Fall, and even after every cutting, the plants should be mulched with a light covering of a mixture of compost, soil and well dried chicken (or other) manure.* See *Sources* at back of this book.

---

*If you are going commercial, you will need to spread a minimum of ten tons of manure and a thin spread of lime as required. A well processed acre should yield about 4,000 plants.

Fortunately the thick fleshy rootstalk is brittle enough to be easily broken into several small parts, the better to create new plants. Or if you're not sure of yourself, you may use a sharp knife to cut the fleshy roots. Certainly most split pieces of root, if planted without delay, will form additional plants. You may try rooting from mid-August through September and October, up to the first frost. Cultivation begun in those three months encourages a healthy establishment of the plants. The first cutting of the foliage will occur in May, a spring planting in August. Prepare a mixture of loose garden soil, a few small stones, and a handful of composted material. Wet the mixture well and insert the root portion. The root cuttings or offsets may be planted 2 1/2 feet apart with the growing points placed just below the soil's surface (as with Rhubarb).

Growing Comfrey indoors will require a little ingenuity but it is worth the effort. In late Fall, preferably on a rainy day, dig out a seedling or as small a plant as possible and transplant it into an oversized pot—about 9–10 inches high. If you use an empty, wooden cheese box as I have, be sure to drill small holes through the bottom wood, to provide proper drainage. On the bottom, spread a layer of inch sized stones or broken pieces of flower pots, onto which add a *mixture* of one part of composted material, four parts of ordinary garden soil, and a handful of small stones. Push the seedling deep into the loosened center of the soil so that the top of the root system is barely visible. Water well. You will have a live Comfrey plant all winter long and an excellent conversation piece.

Make a hedge

You might want to copy an idea presented to me by several Comfrey-enthusiasts. They suggest using this Old-World perennial as a garden plant but only as a two-foot high border or hedge-substitute, for purposes served by similarly intended hedges of Barberry, Box, Privet and Wormwood and Rose varieties.* It serves most admirably, say these enthusiasts, to keep the neighbors' wandering kids and prowling dogs off the enclosed grass or garden area as adequately as any commercially grown hedge. Thus, for all future intents and purposes, the handsome foliage of this richly endowed herb provides more meaningful utility than the colorful tubular corallas. Be sure to remove the flowering stems in the early stages, to better avoid the haphazard growth of future seedlings.

Future cuttings of Comfrey plants are of older plants and the larger-sized leaves, though ill-

*It is interesting that each of these hedge-intended, ornamentals, like Comfrey, offers a variety of worthwhile uses: Barberry—dye, food, and remedy for liver and gall bladder; Box—an alternative and diaphoretic in rheumatic disorders; Privet—an excellent astringent to heal bleeding gums and infected sores, to serve as a gargle and diarrhea remedy; Wormwood—of age-long service as a tonic, as a remedy for fever and jaundice; and Rose—its leaves, flowers and fruits much enjoyed as a food and medicine.

recommended for internal purposes, should be saved for external applications, e.g. for a foot powder, a remedy or poultice for leg sores or ulcers, or as an ingredient of the ever-hungry compost pile or mulch of vegetable hills.

Comfrey in compost

For over thirty-five years I've been an organic gardener. Even earlier, I had stopped using all pesticides in my garden. Compost became my very good friend and an organic garden my close tie-in with today's concern with preventing further pollution and sustaining a healthful environment. "Natural" fertilizer, such as my compost piles, maintains a healthy soil and healthy plants and makes them less susceptible to lice and insects.

The ingredients: hedge trimmings, grass cuttings, weed growths, garden residues, kitchen refuse, market and restaurant "garbage," tree chips and sawdust, fall leaves—even oak leaves and pine needles, swamp muck, fireplace ashes, brown coarse cardboard, old hay, recent corn stalks (cut into 6–inch lengths), all kinds of animal manure, and all the Comfrey leaves and stalks you can spare. Include no animal flesh or bones.

In the preparation of the compost pile be sure to make a thin layer of the Comfrey alternate with layers of other composting material and the soil. A recommendation: Add the large Comfrey leaves as soon as they are plucked, without permitting them to wilt at all. This enables its high content of potassium and nitrogen (and, I presume, its 80 percent moisture content) to be fully taken advantage of. To hasten the composting process, sprinkle with water and pack all ingredients in tightly. Then cover each pile with plastic, to keep the heat in the pile and keep heavy rains out. Because of the ensuing heat, turn the mixture over every 5 days and mix it well. Poke a hole in the heap (to let air reach the organics) and add a little water.

Done in the fall, the compost pile makes you think spring in the dark days of winter. And in early April, you have an ample supply of rich fertilizer for a small-sized garden, to surround your transplants and especially to turn (enriched by more leaves) into the soil. I have already mentioned under *Cultivation*, its application as a mulch for growing produce.

Ever since I've been including Comfrey leaf in my compost piles I've discovered that it is a most speedy activator, a most dependable silent partner, always striving to increase the potency of the compost heap.

· The enterprising Comfrey grower should consider well the herb's usefulness as an ever ready source of green manure, as *instant compost*. All excess of or unused parts of the plant, such as the leaf stalks or plant stems, serve well when mixed into a compost pile or incorporated directly into the soil. To refresh other soil requirements, such as moisture retention, aeration, and co-operation from the friendly worms, as well as supplying an exceedingly high amount of proteins (i.e., nitrogen) and minerals, lay large (fresh) leaves of Comfrey over the walk space between the rows of foods under cultivation and on top of other mulch materials. I have kept this stuff down with heavy cardboard, wide pieces of wood, and the above-mentioned flat stones. Good for such plants that need the wide open spaces—tomato, cabbage, squash, et al. An excellent way to prevent competition from the (then) unwanted weedy nondescripts, the better not to dissipate the soil's nutrients.

Comfrey as fodder

I have already mentioned the wisdom of country folk whose close association with Nature with the soil and weed-like plants and especially farm animals, and whose acute observations of all that transpired in their immediate environs formed their college of formal learning. Which plants do cattle eat, sick or well? Which plants, when convalescing from an illness? Note that many animals will refrain absolutely from all food—they're fasting—until the *whole* system is in complete harmony. How unlike the typical person, with frequent stomach distress, who is continually dosing himself with well advertised or physician-prescribed nostrums which do not, cannot strike at the cause of the gastric trouble. What—go without one meal? Perish the thought!

The observant owner of animals will greatly profit by planting Comfrey wherever possible on their land. This boraginous plant has become a pasture "weed" in the Middle West, where it is relished by grazing cattle. The Russian variety is in greater demand as a fodder crop for pigs. When over a hundred years ago Comfrey became noticed as a fodder plant, various trials showed that a well-cultivated acre of land would yield nearly fifty tons of green food.

Because it is one of the earliest spring arrivals in England, owners of racehorses there make sure the foals are provided with an ample supply of the young leaves as early as March. And since there are four to five cuttings of this productive herb, it has become an important part of diet for young turkeys and other fowl and for chinchillas. "Since I've been feeding Comfrey to my dairy cows, my veterinary bill has been cut in half," stated an Oregon farmer in *Organic Gardening and Farming*. American sheepraisers have now found that when the freshly cut greens are fed to sheep from early spring to butchering time in the fall, those animals raised solely on Comfrey leaves yielded "the most tender, delicious lamb chops," with greater overall saving and at a cost much lower than the market price. Sheep-breeders find that properly cultivated plants may be frequently cut not only to stimulate new growths of young tender leaves but also to provide food for their animals during the winter months since sheep cannot keep up with the growing plants.

Ever since a prickly variety of Comfrey was

introduced in 1811 as a fodder plant into the British Isles from the Caucasian Mountain range of Southeastern Europe, it has been extensively employed as a green food for most animals. This genus, *Symphytum asperimum*, attains the largest height—of five feet or more, with the characteristically prominent prickly stems and large leaves. Since that initial date, various claims for its precious worthiness—as food and medicine for most animals—have been substantiated. Not only did this most inexpensive fodder yield high amounts of flesh-forming essentials; it served both as a preventative and healing agent for foot-and-mouth disease in cattle.

The European farm folk have long witnessed their cattle, pigs and horses eating quantities of leaves, but preferably not as forage. Pigs do eat them in the green state although it takes them a while to get used to the taste. Horses have little appetite for such green herbage even in times of scarcity and then only in small amounts.

Collection and preservation

Gather the leaves before the plant flowers and on the second consecutive dry day. They may be cut every month from June 15 on when the plant is at least a foot high. Take cuttings about 2 to 3 inches above the ground. Leaving sufficient plant stalk prevents damage to the crown, insuring future life to the plant.

Remember, Comfrey grows quite fast and thick and often overruns smaller and weaker plants, unless they are continually cut and the new growths, the seedlings, are transplanted to a more suitable location.

Dry the leaves either by suspending the whole leaf stalks from the ceiling of your attic or cellar; or by placing them loosely next to or over your warm oil burner. If you're using the floor of an unoccupied (but preferably warm) room, be sure to stir the leaves once a day. They may also be dried by suspension in the kitchen hallway or in the garage. In all cases, be sure to keep the drying herbs away from direct sunlight. Above all *don't crowd* the material, to prevent possible mold. Unlike aromatic herbs which are best gathered between 10 A.M. and 12 noon, Comfrey leaves (and roots) may be taken any time of the day providing the leaves are dry.

Collect the roots of fully grown plants in early spring and autumn. The roots are more easily removed on a rainy day—after the rain has stopped. Shake off the excess soil and wash the material (in rain water which should always be collected during a downpour).

Note: Each fall after the roots are collected, it is best to mulch the Comfrey crowns—with a heavy blanket of compost, leaves, or vegetable refuse.

To dry the roots, brush them whole with cold water to remove adhering soil particles, slice them lengthwise and place them on *clean*, rust-free window screens or on trays of cheesecloth or fine mesh wire, or on aluminum trays or baking pans.

To hasten drying, stir or turn the roots every other day. Allow 20 to 25 days to complete the drying process. And remember: Comfrey roots lose between 3/4 to 4/5 of their (water) weight upon drying. Thus ten pounds of freshly gathered roots will end up around two pounds of dried material. However, when storing the dried roots, one must take into account their former fleshy, juicy consistency. The least absorption of moisture invites mold and/or beetles. For root storage, use an air-tight glass jar. Wash it well with hot water and soap and allow to dry thoroughly. Enclose a moth ball or large crystal (naphthalene or paradi-chlorbenzene) in cloth or paper, place on bottom of jar, and cover with cardboard. On top go the cut herbs. Label the container with the name and part of the herb and the date of collection.

Collecting the seeds may present a minor problem. Ordinarily they are gathered when nearing maturity or when the seed pods have opened. But Comfrey's seeds, those brownish-black nutlike kernels which form when the flowers shed their petals, are not easily available. They remain "hidden," Comfreyites tell me, "and pop out when we're not ready for them." One may gently cover the flowering heads either when fresh or as the suspended herbs hang to dry, with cheesecloth or other cloth, or with plastic bags. Or spread out a newspaper under the flowering herb, suspended and drying, and you may collect a few falling seeds. An outdoor seed drop will mean a new plant which is easily transplanted to your favorite spot.

After the third year's gathering of leaves and roots of the *same* Comfrey plants, it is advisable to feed the soil in late Fall not only with the usual composted material but with a thin layer of crushed, ground, or powdered rock (granite, for example) over and around the Comfrey beds. (Do this to your vegetable garden as well.) If you are fortunate enough to live on or near the coast, gather and thoroughly dry all kinds of seaweeds in April or May, cut them up as finely as possible and incorporate them, i.e. dig them well into the soil. Do this also before adding the Fall spread of stone. And last, to put the beds to sleep for the winter, spread a blanket of one-foot lengths of seaweed over the entire flower and garden space and cover with leaves or grass cuttings. An extra: Dig well-rotted manure between the rows when dressing them for winter, spread a layer of compost and leaves, and cover them with a mulch of flat stones. You will be assured of a greater green crop.

Sources

Fresh plants and/or roots:

Gardens of the Blue Ridge, Newland, NC
28657

Gilbertie Herb Gardens, 7 Sylvan Ave., West-
port, CT 06880

Logee's Greenhouses, 55 North Street, Daniel-
son, CT 06239

Meadowbrook Herb Gardens, Wyoming, RI
02898

Nichols Garden Nursery, 1190 North Pacific
Highway, Albany, OR 97321

North Central Comfrey Providers, Box 195,
Glidden, WI 54527

Clyde Robin, P.O. Box 2091, Castro Valley,
CA 94546

Verne Thomas, Hancock, NH 03449

Dried leaves, roots and powdered material:

Herb Products Co., 11012 Magnolia Blvd. North Hollywood, CA 91601

Indiana Botanic Gardens, 626 17th St., Hammond, IN 46325

Kiehl's Pharmacy, 109 Third Ave., New York, NY 10003

Nature's Herb Co., 281 Ellis St., San Francisco, CA 94102

Wide World of Herbs Lts., 11 St. Catherine Street East, Montreal 129, Canada H2X 1K3

Bibliography

(prepared by Ben Charles Harris)

Beach, Wooster, *The American Practice of Medicine*. 1847.

Boericke, Oscar E. *Homeopathic Materia Medicine*. New York: Boericke and Runyon, 1927.

Brown, O. Phelps. *The Compleat Herbalist*. Jersey City, N.J., 1865.

Carter, Annie Burnham. *In An Herb Garden*. New Brunswick, N.J.: Rutgers University Press, 1947.

Clymer, R. Swinburne. *Nature's Healing Agents*. Philadelphia: Dorrance & Co., 1963.

Clark, Linda. *Getting Well Naturally*. New York: Arco Books, 1968.

Coles, William. *The Art of Simpling*. London, 1657.

Curtis, A. *Botanico-Medical Recorder*, Vol. 6, 1838.

Dawes, Frederick A., Ed. *Health From Herbs* (various issues). 100 Portland Road, Worthing, Sussex, England.

Foster, G.B., Ed. *The Herb Grower Magazine*, Vol. 7, No. 1, 1953.

Gerard, John. *The Great Herball*. London, 1597.

Green, Thomas. *Universal Herbal*. London, 1823.

Grieve, Maude. *A Modern Herbal*, Vol. 1. New York: Harcourt, Brace and Co., 1931.

Griffith, Robert E. *Medical Botany*. Philadelphia: Lea and Blanchard, 1847.

Harper, Shove F. *Prescriber and Clinical Repertory of Medicinal Herbs*. London: Homeopathic Publishing Co., 1938.

Harris, Ben Charles. *The Compleat Herbal*. Barre, VT: Barre Publishing Co., 1938.

Eat the Weeds. New Canaan, CT: Keats Publishing, Inc. 1973.

Josselyn, John. *New England Rarities Discovered*. 1672.

Kamm, Minnie Watson. *Old-time Herbs For Northern Gardens*. Boston: Little, Brown and Co., 1938.

Kirschner, H. E. *Nature's Healing Grasses*. Riverside, CA: H.C. White Publications, 1960.

Licata, Vincent. *Comfrey And Chlorophyll*. Santa Ana, CA: Continental Health Research, 1971.

Macalister, Charles J. *An Ancient Medicinal Remedy*. Essex, England: Henry Doubleday Research Assoc., 1936.

Mausert, Otto. *Herbs For Health*. San Francisco: Ben Franklin Press, 1932.

Meyer, Joseph E. *The Herbalist and Herb Doctor*. Hammond, IN: Indiana Botanic Gardens, 1934.

Meyrick, William. *The New Family Herbal*, 1790.

Oliver, F.W. *Natural History of Plants*. New York, NY: Henry Holt and Co., 1895.

Sturtevant, Edward Lewis. *Notes on Edible Plants*, edited by U.P. Hedrick. Albany, NY. J.B. Lyon Co., 1919.

Taylor, Norman. *1001 Questions Answered About Flowers*. New York, NY: Dodd, Mead & Company, 1963.

Usher, George. *A Dictionary Of Plants Used By Man*. New York, NY: Hafner Press, a division of MacMillan Publishing Co., 1972.

Webster, Helen Noyes. *Herbs, How To Grow Them and How To Use Them.* Boston, MA: Charles T. Branford Co., 1939.

Wilkinson, Albert E. *The Flower Encyclopedia and Gardener's Guide.* New York, NY: The New Home Library, 1943.

Wood, Horatio C. and Arthur Osol. *The Dispensatory of the United States of America.* Philadelphia, PA: J.P. Lippincott Co., 1943.

Youngken, Heber W. *Textbook of Pharmacognasy*, 6th edition. Philadelphia, PA: The Blakiston Co., 1936.